AF700628

AUDITING BUSINESS CONTINUITY: GLOBAL BEST PRACTICES

By Rolf von Roessing

Endorsed by
The Business Continuity Institute
www.TheBCI.org

Published by
Rothstein Associates Inc.
www.rothstein.com

ISBN #1-931332-15-0

COPYRIGHT © 2002, Rolf von Roessing.
First Published September, 2002.

ISBN #1-931332-15-0

All rights reserved. No part of this publication may be reproduced, stored in a retrieval system, or transmitted in any form or by any means, electronic, mechanical, photocopying, recording or otherwise, without prior permission of the publishers.

Although great care has been taken in the compilation and preparation of this book to ensure accuracy, the author and publishers cannot under any circumstances accept responsibility for any errors or omissions. Business Continuity is a fast-moving area and inevitably some of the information contained in the book will become out of date. Readers should verify any information contained in this book before making any commitments.

No responsibility is assumed by the Publisher or Author for any injury and/or damage to persons or property as a matter of product liability, negligence or otherwise, or from any use or operation of any methods, products, instructions or ideas contained in the material herein.

Readers should be aware that International, Federal, National, State and Local laws and standards may affect the contents and advice contained in this work, which should be interpreted accordingly. Only the courts can authoritatively interpret the law.

The author welcomes suggestions for inclusion or improvements.

PUBLISHER
Philip Jan Rothstein, FBCI , President
The Rothstein Catalog On Disaster Recovery
Rothstein Associates Inc.
4 Arapaho Road
Brookfield, Connecticut 0608-3104 USA
☎ 203.740.7444
☎ 1-888.ROTHSTEin
203.740.7401 fax
www.rothstein.com
www.DisasterRecoveryBooks.com

info@rothstein.com

THE ROTHSTEIN CATALOG ON DISASTER RECOVERY

TABLE OF CONTENTS

FOREWORD

There are numerous publications that provide a wealth of knowledge about what Business Continuity Management (BCM) is and how it should be done; few offer an explanation of how it can be assessed. Many concentrate on how to develop and maintain a BCM plan; few adopt an holistic approach to BCM and address the key issue of how to develop and maintain a BCM capability based on an understanding of the business and its markets.

This work of Rolf von Roessing is grounded in sound experience and begins to fill the BCM plan/capability gap. It sets out the BCM audit process in a structured and user friendly way that should be basic reading for all BCM professionals and BCM auditors.

A particular acknowledgement is the complexity of a BCM audit and the need for professional BCM expertise as a key element to successfully achieve audit objectives.

The work not only provides a general outline of how to conduct different types of audits but also reinforces their application by providing practical examples and advice to illustrate the step-by-step methodology, including contracts, reports and techniques. The practical application of the methodology enables the professional auditor and BCM practitioner to identify and illustrate the use of good BCM practice whilst demonstrating added value and business resilience.

Dr. David J. Smith MBA LL.B(Hons)
Chairman of the Business Continuity Institute Education Committee

The Business Continuity Institute*'s mission is to promote the art and science of Business Continuity Management world-wide. The BCI promotes the highest standards of professional competence and commercial ethics in the provision, maintenance and services for Business Continuity Management (BCM). It provides an internationally recognised Certification scheme for BCM managers and practitioners. The BCI Professional Recognition Program creates a benchmark for the assessment of best practice in the field. There are now over 1,200 members of the Institute working in 36 countries across the world. Members are drawn from all sectors including Finance, Government, Health, Transport, Retail and Manufacturing.*

For further information, email TheBCI@btinternet.com or visit www.thebci.org

BCI

PREFACE

I was very happy to be asked to write a preface to this welcome addition to the growing library of Business Continuity learning.

Why? As a practicing consultant and trainer of enterprise risk management and business continuity, it has long been a source of discomfort that so many business continuity plans simply pay lip service to real needs. Plans are often over simplistic, over-focused on particular possibilities, ill-considered and incomplete. They make implicit assumptions – about the availability of people, assets and access, for instance – without subjecting those assumptions to challenge.

Around 85% of Business Continuity Plans fail when first tested. Put simply, these plans show fundamental flaws that would have prevented recovery from taking place within the required timescale.

Over 50% of Business Continuity Plans are never tested. This means that those flaws have not been exposed and the plans will almost certainly fail to deliver timely recovery.

These stark figures demonstrate just how misplaced are the hopes of many managers when they rely on such fragile plans. No matter what forethought is given to business continuity management, the actual experience of a disaster bears little relation to the pre-considered events and to plans developed in the relative calm of normal circumstances.

Too often business continuity arrangements are based on specific disaster scenarios and would not withstand scenarios that had not been considered. But disasters are not disciplined. Chaos follows no roadmap. The unthinkable does happen.

It is therefore crucial to businesses that plans are subject to stringent review. That is why I welcome Rolf von Roessing's cogent contribution to this important area. Rolf provides a comprehensive, pragmatic and deeply practical step-by-step guide to Business Continuity audit. I commend it to all who are serious about the topic.

Andrew Hiles FBCI, MBCS
Oxford, UK
June 2002

Andrew Hiles was founder and, for some 14 years, Chair of an international business continuity user group. He was a founding director of the Business Continuity Institute and a founder of the World Food Safety Organization. An international consultant, trainer and coach in enterprise risk management and business continuity, Andrew has some 25 years experience in this area and is a director of consulting group Kingswell International.

INTRODUCTION

This book presents a general methodology and a framework for auditing Business Continuity Management (BCM). The main purpose is to provide a single work of reference for auditors, managers working in business continuity and consultants.

BCM is a complex field. It covers business issues and technology with a perspective on the entire enterprise. The business continuity manager, and the auditor, require a diversified set of skills and extensive knowledge to assess business continuity as a question of business survival. There has been a lot of confusion about the terms "business continuity," "disaster recovery," "IT security" and many other words attempting to describe the continuation of critical business processes under adverse circumstances. However, for the auditor these terms refer to one and the same notion: businesses should take adequate precautions to ensure that no going concern issues arise from crises or disasters.

Some companies decide to take a cautious stance with regard to continuing their operations come what may: they prefer to "err on the safe side" and rely on preventative measures. Other firms, perhaps in an industry where "speed to market" and competitive pressure require a faster pace, may prefer to reduce investments on prevention, while putting in place a robust crisis and disaster management mechanism. Both types of corporations nevertheless pursue the overall goal of business continuity, by either avoiding risks or disasters (if they can), or by making sure they can deal with these events.

In a sense, BCM means "reading the future" or trying to safeguard an organization against unforeseen events. Management is still forced to address precisely this issue, by carefully evaluating their options and then making an entrepreneurial decision about the acceptable level of remaining risk. To the auditor, it is important to understand how this decision has been reached and whether it can be justified from a financial, operational and managerial point of view. Neither the overly cautious nor the reckless manager will succeed in today's market – the BCM auditor should provide a sounding board and an objective business partnership to the management of the company being reviewed.

BCM audit is therefore an important element of ensuring corporate survival. The audit result incorporates issues of compliance, highlights weaknesses and provides reasonable recommendations to management, whose experience may be enhanced and improved by the auditor's objective input from other corporations or industries. It is not to be confused with the much narrower field of IT audit. This book has been deliberately restricted to business continuity rather than IT continuity to highlight the all-important differences between the two.

The contents have been arranged around the Business Continuity Institute (BCI) / Disaster Recovery Institute International (DRII) Professional Practices for business continuity as well as other standards such as CobIT or ISO / IEC 17799. Some elements may look familiar to the experienced auditor who may still benefit from using this book as a reference manual or as an instructive tool for groups of auditors. This is intentional, as BCM and related audit questions should "fit in" with tools and models that are recognized and proven in the field.

HOW TO USE THIS BOOK

This book is a toolset to assist you in planning, conducting and documenting a review of the business continuity management (BCM) process within a company or institution. It is structured in three main sections. The first part explains how to plan an audit from beginning to end. The second part contains a full audit program that you may use at varying levels of detail to support your audit strategy and plan. The third part contains samples of an audit report and selected work papers to help you put the plan and program into practice.

If you are a financial auditor, or an internal auditor tasked with reviewing business continuity, this may be a new field to you. Likewise, if you are a business continuity manager who has been assigned the task of being an auditor, this is a new way of looking at BCM, rather than implementing it. **Chapter 1** explains the concepts of BCM and audit seen together. It shows how to formulate the framework and scope of a BCM audit, how to define audit plans and how to write a clear and concise audit program that management and other stakeholders will understand and buy into.

As an auditor, you are managing the practical phase of a BCM review. **Chapter 2** explains how to schedule the review, how to estimate time and effort, and how to streamline the process of formal audit steps. Known difficulties and pitfalls, many of them unique to BCM, are explained in detail. Even if you are a seasoned audit professional, this chapter may help you in identifying typical problems associated with reviewing a complex process and interacting with a wide range of managerial and technical responders. As a business continuity manager, Chapter 2 may help you understand the challenges presented by reviewing the BCM concepts without actually managing them yourself.

Chapter 3 outlines methods of analysis that you can use to arrive at a well-founded audit opinion. As a financial or internal auditor, this chapter will allow you to evaluate your findings and to avoid time-consuming detail when reviewing the BCM process. As a business continuity manager, you will find Chapter 3 a useful tool for looking at any given part of a BCM process and for comparing findings against your own experience and best practices.

The success of your work as a BCM auditor depends on clear, concise audit reports that are easily understood by management. **Chapter 4** explains how audit reports are structured, written and presented to your stakeholders. In this chapter, you will find samples and templates ranging from small, detailed reports to a large set of reports designed for an international BCM audit.

Section 2 is a standardized audit program divided into work areas. You will find detailed audit questions covering all aspects of business continuity management. In the course of your BCM audit, you can use parts or the whole of the standardized questions for your audit plan and program. The standardized audit program is designed to give you additional information on risk ratings, recognized standards and additional materials that you may use to understand each item, as well as to communicate it to audit teams or the auditee organization.

For each item within the standardized audit program, the legal, regulatory and technical background is explained in detail. Detailed audit steps have been included for each question to give you indications as to the time and effort required during the audit. Suggested standard wordings for findings and recommendations have also been included.

Work area 11 contains detailed audit instructions for some national jurisdictions where different rules may apply. You can use these to guide your audit teams, and to find out what materials you may need to understand and evaluate when reviewing BCM abroad. The national parts of area 11 include the Central and Eastern European world to give you an overview of what to look for even if a foreign language is used.

Work area 12 will support you when reviewing typical BCM software tools. You will find useful hints and technical references to give you quick access to typical problems and difficulties that may constitute important audit findings.

Section 3 contains a sample audit report that is based on the examples used in Section 1. Selected work papers have been added to provide an indication as to the ways in which you might use the standardized audit program.

Depending on your previous experience with audit and BCM, you can use this book as a reference work or as a step-by-step guide for hands-on project work. However, it is *not* a "one-size-fits-all" guide along the lines of "BCM-in-a-box for $ 9.99." Whether you are a novice auditor or a seasoned BCM professional, it is likely that you will use the book in different ways. Your comments on using it, and suggestions for improving the framework suggested, are welcome.

SECTION I

AUDIT GUIDELINES FOR BUSINESS CONTINUITY MANAGEMENT

1

AUDIT FRAMEWORK, SCOPE AND PLANNING

If you are a financial auditor, or an internal auditor tasked with reviewing business continuity, this may be a new field to you. Likewise, if you are a business continuity manager who has been assigned the task of being an auditor, this is a new way of looking at BCM, rather than implementing it. Chapter 1 explains the concepts of BCM and audit seen together. It shows how to formulate the framework and scope of a BCM audit, how to define audit plans and how to write a clear and concise audit program that management and other stakeholders will understand and buy into.

Introduction

Conducting an audit of business continuity planning and management presents an unusual challenge to financial or IT auditors. While audit automation, CAATs[1] and other acronyms have found their way into traditional audit projects, the more technical fields of IT, facilities and business continuity have yet to develop a similar strategy for facilitating detailed audit steps. More often than not, it is even difficult to determine the scope of a review in terms of time and effort, considering that the

[1] Computer Assisted Audit Tools

knowledge required to do so is diverse and not easily obtained.

More importantly, if the business continuity review is to take place within an annual financial audit program, there is often no room for budgetary, or indeed any other uncertainty with regard to the delivery of results. The auditor thus faces a challenging and complex task. Business continuity, as a local or regional activity, almost always requires the direct involvement of experts who can assist with technology, local regulations and interpretation of factual findings. Unfortunately, the number of experts is often limited, and published audit guides and programs are rare. The BCM auditor is therefore a project manager as well as an expert in the field, having to strike a balance between the task at hand and the overall constraints imposed by commercial thinking.

For the business continuity expert, on the other hand, the general world of audit thinking may be an alien one. Designing a business continuity management process is a very different task from scrutinizing an existing set of plans and procedures, verifying them against predetermined standards, and delivering an audit opinion in a structured manner. To the BCM specialist, the main aim may be to "make it work," while the auditor must examine quite a significant number of other aspects.

The following sections provide an outline of a model business continuity audit from inception to planning to final report delivery and "post mortem" analyses. Depending upon the overall objective of the audit at hand, different steps and different levels of detail may be appropriate. For instance, a financial auditor wishing to establish the presence of corporate risk management may see business continuity management as a subset thereof, whereas a consultant reviewing existing continuity plans prior to a large reorganization project might prefer an in-depth audit of all continuity-related issues. The emphasis given to individual audit objectives or areas of business continuity may therefore vary considerably. A number of trade-offs must be taken into account, for instance the budgetary and time constraints usually imposed on any commercial or third party audit project.

Consider the following two examples which give a fair notion of BCM audit as opposed to operational BCM. They also illustrate how audit and expert staff might view the task at hand, and thus help to avoid misunderstandings and confusion as to the overall motive for conducting an audit.

> ***Example 1***: *A large organization has recognized the need for business continuity as a management process. Senior management decide on an internal structure, comprised of a project group, a BCM manager and several junior team members. It is further decided to bring in a number of external consultants who are experts in BCM. All group members agree on the fact that BCM in this organization is a "greenfield" exercise, and that no business continuity steps have been taken beforehand. Hence, the first task is to develop an overall framework. Subsequently, project plans and detailed milestones are put in place, as are budgets for the project group and the external consultants. The BC manager and project leader may have decided to use an acknowledged standard for designing the BCM architecture and for monitoring departmental or business process planning.*
>
> *Some time later, the BCM effort is well under way. While both the project group and corporate departments are busy creating their plans, assigning roles and responsibilities, and reporting on progress and budget issues, the level of maturity is still fairly low. The pragmatic approach taken by the BC manager is a useful one, as it allows departments to gradually aspire to perfection and completion of their planning. In a step-by-step process, the organization is likely to reach a satisfactory*

level of BCM maturity within a year or two.

The BC manager's perspective is to design, deploy and monitor this planning project. It is usually a major task with a significant budget. Managerial success therefore hinges on traditional project management skills – deliver the business continuity product on time and on budget while managing diversified corporate interests and stakeholders – and, the ability to convey sound thinking and pragmatic objectives that should be accomplished by the overall organization. Most importantly, both novice and experienced BC managers know that their world is less than ideal, and that planning and project management imperfections will inevitably occur. More often than not, political and financial constraints are put on BC projects. Group or team decisions may blur the clear-cut, theoretical BC standards originally applied to the project. Senior management may decide to intervene for strategic reasons or as a result of far-reaching corporate cost-cutting decisions.

Example 2: *A similar large organization conducts a straightforward year-end financial audit, using the services of a well-known international accounting firm. Senior management are informed that, to their surprise, disaster recovery and business continuity plans do not exist in the divisions reviewed. The audit report contains recommendations that – if implemented word for word – would entail far-reaching and complex BCM projects, not to mention the enormous budget needed for such an endeavor. To reinforce this latter point, the auditors have presented a proposal for consulting services that is worth several hundred person-days and will create significant restructuring activities throughout the organization.*

At the final audit meeting, senior managers are faced with the choice of pursuing the matter or having to live with a significant audit finding that might even end up as a remark in the official statement. The auditor presents laws, standards and industry initiatives suggesting that BCM is on the brink of becoming a mandatory part of day-to-day business. Comparing the current situation to an ideal world, the auditor further suggests that a lack of controls and monitoring might lead to substantial damage, or even litigation and lengthy liability issues in the event of a disaster.

Senior management, technically-oriented middle managers and specialists accept these findings with a grain of salt: it is known that perfection in management is very rare indeed. It is also known that the risks and events depicted by the auditor are quite improbable. Historic events leading to damage have been studied, and appropriate prevention programs are in operation. Fire protection, earthquake protection and suitable procedures for bomb threats are in place. The organization, after prolonged discussions, feels that trying to protect itself against each and every conceivable (and unknown) risk is not feasible and too expensive. It is however acknowledged that some weaknesses exist, probably in the area of IT security where new threats emerge every day.

The auditor's perspective is one of impartial review against defined standards. Even if the BCM specialists on the audit team are well aware of the usual shortcomings and organizational difficulties existing in day-to-day management, they have no choice but to highlight any deviation from their "ideal world" which is an integral part of the idea of auditing. Not mentioning the imperfections in BCM, or showing a wide-ranging understanding of the managerial problems and budget constraints within the organization, might be a realistic approach. However, the audit mandate and its intrinsic notion of leaving no stone unturned suggests otherwise. Not mentioning a finding, or allowing for anything "less than perfect" might create a liability case for the auditor. Non-compliance must not be condoned, as every auditor will learn when training for CPA[2]*, CISA*[3]*, or other audit-related certifications and studying the code of ethics.*

The two examples show how BCM might be perceived differently, depending upon the managerial position and its responsibilities. It is perhaps not surprising that BC managers and BC auditors arrive at the scene of audit from very different starting points. It should be borne in mind, however, that the BC auditor may well have come from a background of operational BCM experience, and the BC manager may have been an auditor at some point. Consequently, the review of business continuity management does not normally represent an overall criticism directed against the managers and specialist teams working for the auditee organization. While these individuals may feel that they have been unfairly criticized, given their internal constraints and priorities, the auditor may feel misunderstood in their approach. The seemingly harsh wording of findings and audit opinions creates differences of opinion which might be easily resolved if both parties could agree on a common ground for discussion.

It is suggested here that both the auditor and the BC manager – perhaps unconsciously – pursue the same goals and objectives: to create, implement and maintain a flawless process of business continuity management. While there are impediments on the organizational side and seemingly exaggerated demands from the audit side, the desired end result for both parties is probably identical. The following sections of this book are based upon this assumption.

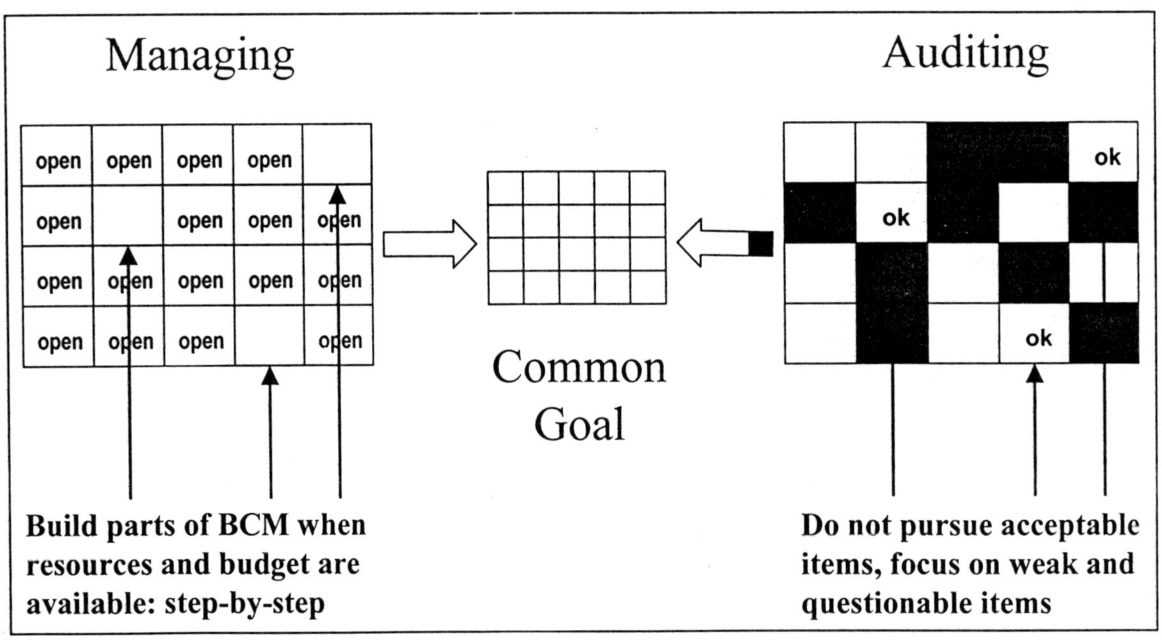

---

[2] Certified Public Accountant

[3] Certified Information Systems Auditor

Audit Framework

BC audits are usually conducted within the larger framework of either an audit review – whether financial or technical in nature – or a preliminary activity to a consultancy assignment or internal BC project. The worst case may even lead to an investigative audit, of a legal and/or actuarial nature, in the aftermath of a disastrous event. The audit function in its traditional sense is therefore less of a self-serving purpose, and more of an assistance function serving a particular management objective or set of objectives:

- Compliance (with internal management guidelines, with external regulations or standards, etc.)
- Project preparation, project control (ascertain the status quo prior to initiating a project, or during a project)
- Feasibility study (similar to project planning – it may be advisable to have an audit performed beforehand to see if the organization is capable of sustaining a large project)
- Due diligence and other major strategic objectives
- Forensic investigation (securing evidence, matching plans against actual events...)
- Others...

The list clearly indicates that the audit framework must be known in order to establish clear and concise objectives. These in turn will determine the scope of the audit itself, such as level of detail, timeframe, budget, and results to be reported. Hence, it is essential for the auditor to clearly delineate the framework prior to commencing detailed planning. Some standardized frameworks have been suggested for other audit applications, such as finance[4] and general IT[5]. If applied to an external compliance audit or periodic internal audit, these standards can provide invaluable insights into the management objectives as well as any predetermined audit scope or mandatory reporting. Despite these indicators, frameworks are typically restricted to providing direction rather than detailed guidance as to audit steps. Business continuity is perceived and described as one of many points included in the overall review, having parity with others such as information security or financial records. In selecting, setting and defining the framework, senior management can influence the priority assigned to business continuity and the related audit steps.

Some examples given below show how standards might be used to justify the need for business continuity management.

The CobiT model – as a set of control objectives for information technology and its organizational integration – is subdivided into four broad categories of controls. While it is originally designed to cover all IT-related processes within an organization, the interesting part with regard to BCM is found in "planning and organization" and "delivery and support." The former contains a control objective "risk management" (PO 9), whereas the latter is concerned with "ensuring continuous operations" (DS 4). Both are clearly aspects of BCM, but they are usually addressed in the context of an IT audit. As will be shown below, this is too restricted and may be pointing in the wrong

[4] Several standard audit frameworks, mainly for financial and year-end reviews, have been put forward by the so-called "Big 5" accounting firms, for instance the "Audit Innovation" and "Global Release" methodologies at Ernst & Young LLP.
[5] CobiT – Control Objectives for Information and Related Technologies – issued by the Information Systems Audit and Control Association (ISACA) is a typical example of an IT-based audit framework which, incidentally, addresses aspects of business continuity, IT disaster recovery and risk management by putting them into context. See also CONCT – Control Objectives for Net Centric Technology.

direction: both auditor and manager might think that having appropriate controls under the above-mentioned headings is equal to having good business continuity management in place. This may be true for the IT environment and its technology, but CobiT does not address the more general concerns of the business as a commercial enterprise.

The former British Standard BS 7799, now known as ISO / IEC 17799 (2000)[6], is a comprehensive IT security standard that deals with a wide range of known threats and control weaknesses. Since 1999, the standards contain a full chapter on "business continuity planning" demanding a reasonable BCM process within the organization. While this is a direct reference to BCM that the auditor can use to justify looking at weaknesses and controls, it is still somewhat restricted. The overall objective of ISO 17799 is security, not continuity. It is therefore likely that the audit will look at BCM at a high level without addressing processes outside IT and general security. Clearly, risks and threats that are not related to the notion of "security" will be missed when using such a narrow approach.

The second version of the Basel Accord, which has come to be known as "Basel II," contains a number of references[7] to controls in business continuity management. Again, this might be used by the auditor to make BCM fit into the framework of a review in the banking sector. Unfortunately, the Basel II approach is primarily concerned with banks and their risks. The "operational risk" in banking terms contains IT-related matters, other control failures and the lack of continuity. The Basel approach is a good example of a specific sector defining its own priorities and guidelines. However, it might be difficult to use the banking view of operational risks in other industries.

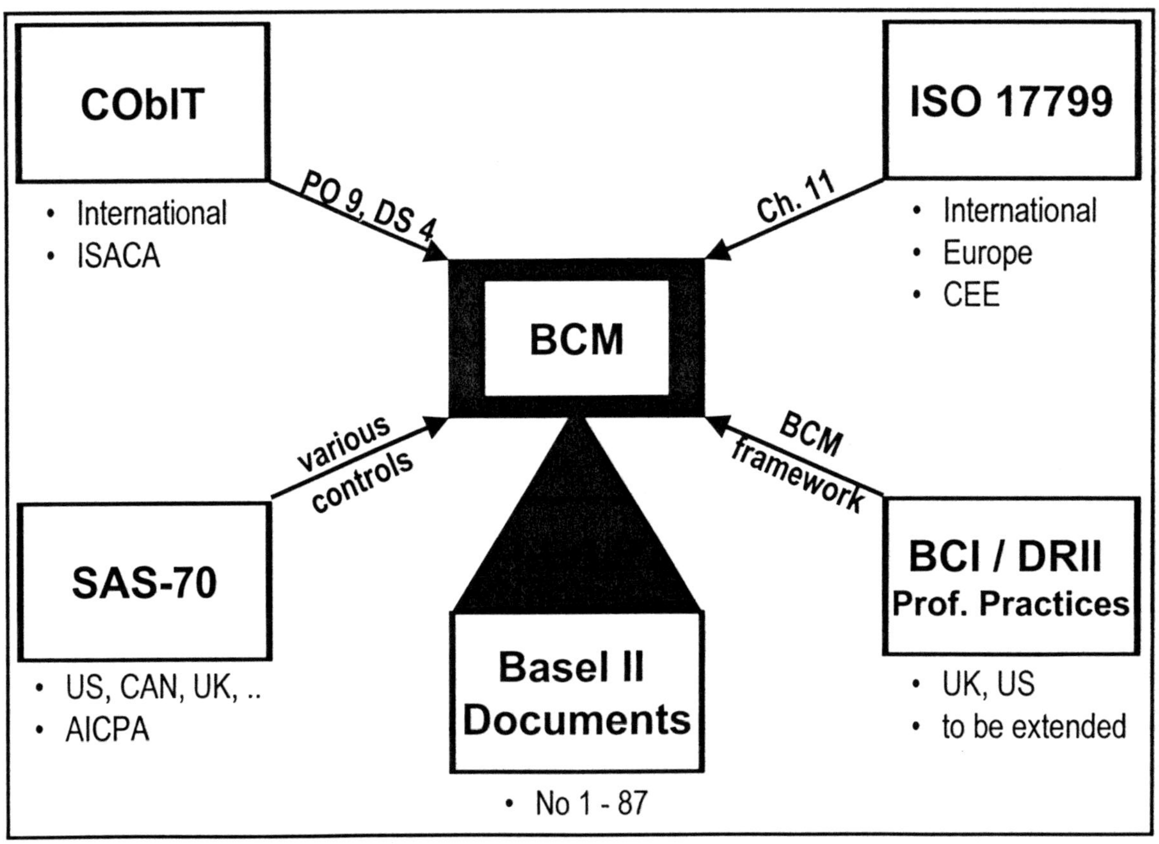

[6] London: The British Standards Institution (BSI), 2000.

[7] the sources are all available from the Bank for International Settlements at www.biz.org. Some of the papers are consultative in nature, whereas others provide so-called "principles" or "sound practices" for specific risks and banking problems.

As an auditor, one should regard these examples as what they are: a toolbox with different implements that may be appropriate to use in certain situations. When reviewing a bank, the inclusion of "Basel II" requirements for operational risk and BCM is likely to create additional value. An organization using ISO 17799 or BS 7799 might benefit from a link between the BCM audit and their broader standardization initiatives. A complex IT environment with CobiT-style processes will greatly benefit from clear audit statements directed at "PO 9" or "DS 4" respectively: the IT managers of that organization will understand where and how to improve, as the CobiT world is much more familiar to them than the idea of auditing BCM as a management process.

Consequently, these standards and frameworks should become part of the auditor's toolset for reviewing and reporting on business continuity. There is a considerable degree of overlap between CobiT, ISO 17799, Basel II and even some US accounting standards and directives such as SAS 70 (see below). "Killing several birds with one stone" is one of the benefits that may be derived from using the toolbox with ease. Another advantage is the increased level of support that can be leveraged from the departments or managers championing the projects related to these standards: a BCM auditor endorsing the ongoing CobiT process modeling project will be seen much more favorably by the CobiT project manager.

Using other standards and highlighting the common ground they share with BCM is similar to using different languages to communicate. Being able to "speak" in the language of, for instance, ISO 17799 will almost always prove more effective than having to stick to a set terminology of BCM that requires translation to other "languages," i.e. frameworks. Hence, the examples discussed above will be used throughout the remainder of this book to point out similarities and common areas of interest.

Audit Scope

Once the organizational framework of the audit is known, some constraints and limitations with regard to the scope will have been determined. The sum of all activities and steps relating to the audit may now be defined as a finite project, with quality objectives, resource assignments and a reasonably well defined timeframe[8]. For the purpose of determining the scope of the audit, it is advisable to apply a life cycle model of business continuity management[9]:

- Analysis phase (risk analysis, business impact assessment)
- Strategy phase (definition of overall BC strategy, senior management decision making, cost-benefit analysis)
- Detailed implementation phase
- Test phase
- Maintenance phase

---

[8] The PMBOK methodology published by the Project Management Institute (PMI) is recommended as a framework for managing single or multiple projects.

[9] This assumes (correctly) that business continuity within the corporate environment should be treated as an ongoing process rather than a one-off project.

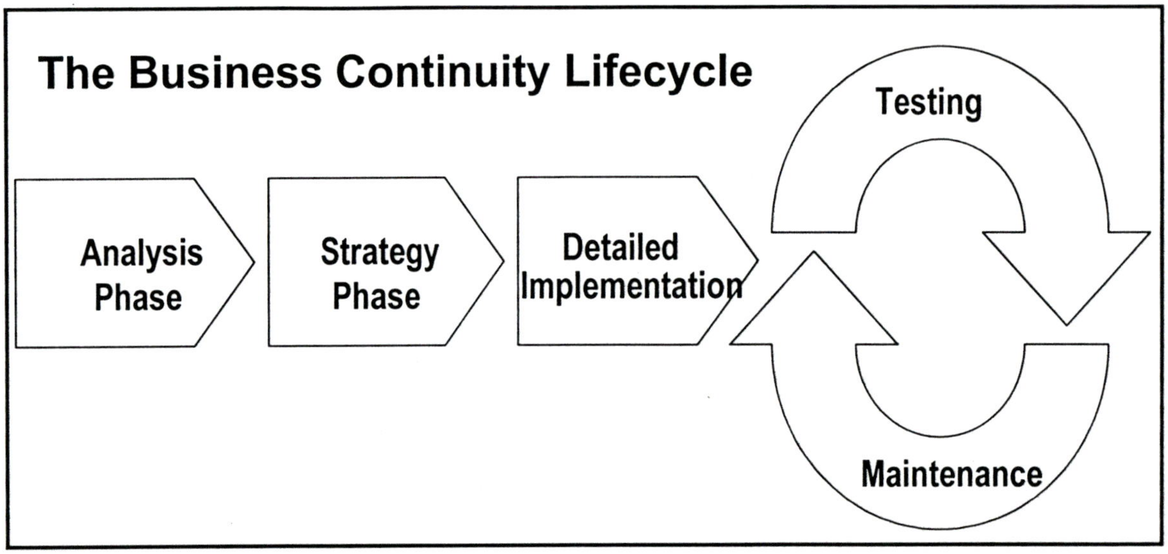

The first two phases typically represent first-time planning within an organization. A fundamental approach to risk management – instituting the methodology, identifying risk in broad terms, categorizing processes, determining process criticality – is usually performed when there is no evidence of prior business continuity initiatives. Analysis and subsequent definition of the overall strategy may be equated to a "kick start" mechanism, thus preparing the corporate environment for the introduction of an innovative process or method. Phases 3, 4 and 5 form the circle of business continuity management: detailed planning is succeeded by testing of plans, hypotheses and detailed instructions, whereas the fifth phase implements lessons learned and necessary improvements to the plan. The need for improvements will in turn lead to revisiting the detailed planning phase.

In auditors′ terms, the five phases of business continuity resemble the introduction of a new ERP[10] package. If no such package is in place, a proportion of the project will necessarily be devoted to setting an overall managerial framework. Activities such as drafting the accounts, journals and cost centers would fall into this category. Subsequent steps would identify areas requiring detailed internal controls, such as payroll or (in a bank) trading floor support. Undoubtedly, such specific areas would be regarded as critical in comparison with others. Once this framework is established, detailed implementation plans would be drafted for individual modules of the ERP package and their customization for different departments or groups. Similar to the fourth and fifth phase of the business continuity life cycle, the ERP package would then be tested in its initial setup, and maintenance of the parts not performing to standard would take place.

Business continuity as a concept lends itself to precisely the same audit steps. Initially, the overall model and approach will be defined by senior management. Then, the current processes and departmental activities will be examined in detail, with a view to establishing the critical (vital) processes. It is no coincidence that departments with "critical" ERP applications would also be considered critical in terms of continuing their operation in the event of a disaster. Referring to the above-mentioned example of payroll, the department would require both the ERP system and – adequate – other resources to ensure timely payments to employees. However, it should be noted that the ERP system alone cannot support the department's activities.

The identification, evaluation and subsequent listing of all things "critical" or "vital" to the survival

[10] Enterprise resource planning, for instance SAP, JD Edwards, Peoplesoft and many others.

of the enterprise as a whole are the true reason for conducting a BCM audit: continuity management is reviewed because the auditor should render a true and fair view of the organization's ability to survive under adverse circumstances. If this ability is impaired, the assumption that the enterprise is a going concern must be questioned and validated much more thoroughly than would be the case if BCM were evident and up to date: imagine two companies hit by a major disaster. One company has made preparations for running its core operations in "emergency mode" for a limited period of time. The damage caused by the disaster is limited, and after a few weeks, business is back to normal. The other company, having neglected the preparations, suffers severe damage and is forced to file for chapter 11 bankruptcy. Subsequently, it is taken over by a competitor. This simple example illustrates how the "going concern" assumption is quite closely related to BCM, although this issue is still being debated in financial audit circles.

The second phase of the life cycle model – strategy formulation – is based upon the findings of the preceding phases. Typically, the business continuity manager or consultant will identify broad strategic options, such as "do nothing," "calculated risk" or "maximum coverage" and quantify them in terms of cost and expected benefits. The comparison of such options is often presented to senior management as a synopsis, or decision matrix outlining the "for" and "against" aspects of each option. The outcome of the strategy phase sets the context for detailed implementation and the introduction of internal controls for business continuity. Once management decisions have been made, plans can be measured against specific objectives and expected results[11].

Clearly, business managers seek to minimize the cost of continuity. BCM is a cost factor just like any other item on the profit and loss account. It is a legitimate concern to spend as much as is required whilst avoiding excessive spending on continuity. In terms of strategy, the risks and threats, the probability of occurrence, and the expected impact on the organization have to be evaluated and approved by management. It is up to the entrepreneurial judgment of the officers and directors to determine how much residual risk – after implementing preventative steps, controls and procedures – they are prepared to tolerate. This "appetite for risk" is the decisive factor in opting for a more limited BCM strategy as opposed to a comprehensive "safety net". This is not an audit decision. Corporate officers are free to decide on the level of BCM, as they would be free to decide on insurance policies or bonding. However, it is the auditor's duty to highlight possible consequences and impacts of the strategic decision taken by management, and to clearly identify liabilities and other threats related to the "going concern" question.

The managerial framework thus established is disseminated throughout the organization in the three remaining phases of the business continuity life cycle. Detailed planning, usually at departmental or group level, will ensue, and testing and maintenance of individual plans and controls will be subject to central monitoring. This process is naturally iterative, as the business will inevitably undergo changes over time. Any major change, such as a merger, IPO or restructuring, must be reflected in the corresponding arrangements for business continuity, i.e. the continuation of processes in question under adverse circumstances.

For the auditor, each phase of the business continuity life cycle requires a different approach. Depending upon the maturity of the entity reviewed, the first step is to determine which of the stages have been completed. If the company happens to be in the early stages of a business continuity project, the scope of the audit can be reduced to existing plans and documentation. If, on the other hand, a company has reached a level of near perfection after several iterations of the life

[11] Similar thoughts would apply to service level agreements with third parties. When considering outsourced functions and benchmarks for availability and contractual performance, it is apparent that definitive management decisions are needed to monitor the relationship.

cycle – and, perhaps, some critical events – the audit would have to be much more comprehensive to adequately reflect the status quo. In an approximation of capability maturity models[12], the following list may be indicative of stages that the entity reviewed may be undergoing at the time of audit:

- **N/A**: business continuity has neither been discussed nor addressed in a managerial sense. No documentation exists. Some isolated initiatives for managing specific risks (fire, flooding etc.) may have been completed following past events or incidents. However, the organization as a whole would not classify these as "disaster recovery" or "business continuity". Typically, the specialist solutions implemented for some specific risks and threats are not coordinated. People within the organization are not aware of the risk management steps taken, and only a few individuals such as facilities managers or IT staff have any knowledge of the "hidden" precautions taken.
- **Novice**: the entity under review is vaguely aware of business continuity, perhaps indirectly in the sense of IT disaster recovery. No attempts have been made to categorize business processes or extend the paradigm of IT recovery to core business activities. Typically, some middle managers within the organization will have recognized the need for mitigating risks and threats, without calling it "business continuity." Some departments may have tried to implement initial steps, but it is not uncommon to find that such initiatives have been discouraged by higher levels of management due to the burden they might place on departmental budgets.
- **Intermediate**: technical recovery has been initiated and partially implemented. Core business processes are not covered by the methodology applied. Management uneasiness is evident, but no steps have been taken. Organizations at this stage of maturity often display a comprehensive documentation on IT disaster recovery, decentralized backups, contracts for alternative data center sites, and elaborate technical protection concepts. Typically, these solutions have been implemented to mitigate "security" risks. IT recovery, security and other terms are used synonymously with business continuity, but it is often found that these expensive technical precautions are seen as sufficient to protect core business. When the organization is close to reaching the "advanced" level of maturity, it is likely that there will be quite a few managers acknowledging that more comprehensive plans are needed. Last but not least, there may be elaborate "Y2K" plans that are readily available on request, showing a good understanding of business continuity.
- **Advanced**: business continuity has been addressed in one or more projects. Some plans exist, and they may be comprehensive. Business continuity, as a notion, is integrated by means of a part-time or full-time position within the organization. This stage is often characterized by a "top-heavy" BCM process: headquarters and some large subsidiaries may have had a head start, perhaps assisted by local or regional regulations. At HQ, the teams and BC managers are experienced individuals with a clear sense of direction and a willingness to expand BCM to other parts of the corporate network. Other subsidiaries and recently acquired entities are far less developed in their BCM efforts. There may have been cultural, budgetary or other difficulties that have prevented

[12] cf. Paulke (1995) for the Carnegie Mellon University maturity model for software development. For the purposes of this book, the notions of "ad hoc," "repeatable" and so forth have been replaced by more tangible criteria reflecting typical maturity levels found in business continuity management.

standardization and international rollout. When discussing BCM, the majority of managers and departments will have adopted continuity-based thinking, and the reasons for not having implemented this throughout the organization will be found in other areas such as finance or even language.

- **Mature**: business continuity management exists as an ongoing process. Plans, concepts and documentation are well developed and up to date. Insular weaknesses in BC may exist. Typically, senior managers have demonstrated their "buy-in" to BCM by placing responsibility for it with an officer or senior vice president. The dynamic nature of the continuity process within the organization is clearly visible, and frequent testing and maintenance activity can be observed. Another indicator may be the ongoing and active involvement of managers in BCM-related institutions and working groups. Characteristically, a standardized set of documents will exist which is distributed or even published beyond the borders of the organization.

It is obvious that the maturity stage will determine audit scope to a large extent. For instance, a company displaying signs of being "untouched" need not be audited in detail. The mere statement of non-existence of business continuity will suffice to escalate the matter. Conversely, a company in the "mature" stage will benefit only if the auditor can provide a meaningful, targeted contribution by identifying insular weaknesses or potential improvements at a detailed level.

Step 1: Determine the BC maturity stage of the entity under review.

Once the maturity stage of business continuity has been determined, more detailed objectives should be set. Referring to the above examples, terms such as "vaguely aware" or "partially implemented" do not in themselves present a quantifiable result. Moreover, the fundamentals of both financial and information systems audits, may not grasp the problems inherent in continuity management. For instance, "completeness" as in completeness of a transaction log, cannot be applied to a continuity plan until such a plan has been put to the test. Hence, a different set of criteria is needed. These will often depend upon the overall idea behind the audit.

The Compliance Audit

A typical compliance audit might measure facts and evidence against some accepted standard(s), at a defined level of detail. Compliance is often tested as a subset of annual financial or IT audits, where business continuity would be grouped as part of risk management[13] or part of continuous IT operations. In a broader sense, compliance with external regulations may extend the scope of the audit, as was the case with specific reviews preceding the "Y2K" date change[14]. However, the general focus is on measuring compliance with, or comparison with, specific standards and defined expectations. These may have a fundamental impact on the validity of the business continuity project reviewed. In many cases, business continuity plans will not stand if certain regulations have not been adhered to in formulating the plans. As a result, elements of compliance audit will usually be present in an audit program, whether they be an "*Appropriate / Inadequate?*" or "*How much work needed to accomplish an appropriate level of compliance?*" question.

[13] cf. the "Planning and Organization" and "Delivery and Support" control domains within CobiT, op.cit.

[14] Compliance and publication requirements as per Y2K Readiness Disclosure Act

Examples of compliance audits are often found when BCM is reviewed as part of a broader audit scope. For instance, a security review in the context of an ISO / IEC 17799 audit (formerly known as BS 7799) will address the requirements for business continuity planning set down in section 11 of the ISO standard. Other compliance issues might arise when an organization is reviewed under SAS 70 (data centers) or the ISACA CobiT framework for IT audits. The fairly detailed questions on business continuity may appear somewhat misplaced at first sight, but their validity is often explained by the overall motive behind the standard, as shown in the following examples:

- **ISO / IEC 17799**: while the standard itself is concerned with IT security, the very high probability of so-called "denial of service" attacks, or the exploitation of unplanned systems outages, is a clear security risk. If an attacker knew that by executing a denial of service against critical systems, the whole business could suffer considerable damage, this would undoubtedly lead the attacker to focus on just that to maximize the impact. Systems and IT-dependent core business processes that are protected by a solid business continuity process are obviously better protected against security breaches of the "crash" type, thus reducing the overall security risk.
- **SAS 70**: when reviewing a data center, there may be service level agreements for providing vital services to third parties, for instance to banks and financial institutions. Any unplanned outage or systems downtime might have a disastrous impact on the customer and subsequently on the data center. Liability issues and possible litigation clearly indicate that sound business continuity practices should have been adopted to reduce this risk. Given that the data center organization is more than a technical shop, the managerial aspects of running the data center should also be covered by the BCM program.
- **CobiT**: the audit of IT systems and applications represents just one side of the equation. The control objective "DS4: Ensuring Continuous Operations" is quite similar to the requirement of the SAS 70 example (above). However, the control objective "PO9: Risk Management" addresses the broader context of risks emanating from IT which might endanger the overall business. In typical e-business environments where a large part of the business depends on IT processes, the lack of continuity in IT operations might cause considerable damage to the organization. Likewise, the business management side might be unable to deal with IT-related failures if senior management is not equipped with a solid business continuity plan that explains what to do in an IT-related emergency (consider public relations, financial issues etc.)

Preparatory Audits

Preparatory audits – as a prerequisite to initiating a business continuity project – ascertain the status quo, with a view to identifying the amount of work or effort needed to bring business continuity management to the desired level[15]. The important distinction between this type of audit activity and the aforementioned compliance review is that both auditor and auditee will be aware of a gap between the existing state of planning and the future state of continuity management. In line with the maturity stages listed above, a preparatory audit would normally seek to establish the "distance" between the status quo and the next stages of maturity. Quite apart from the sometimes rather mechanistic set of findings, potential risks and standardized recommendations, reporting for a

[15] cf. the BCI (Business Continuity Institute – www.theBCI.org) evaluation toolset which may be used to define preparedness at a high level. However, it should be noted that this is a *survey tool* rather than an audit programme.

preparatory audit may even produce a proper project plan, or a feasibility study. It is important to note that, for instance as a consultant, the auditor does not merely address weaknesses with a wary eye on audit independence. On the contrary, *the fact that an overall decision to pursue business continuity management has been made (presumably) should encourage the auditor to identify steps to be taken rather than criticize what is missing.*

> Consider an organization that has experienced a change of direction, as a new CEO has taken over. The new management team may have identified weaknesses in business continuity, probably by reading the previous year's financial audit report. The CEO's background is likely to encompass experience with BCM in other companies, and the status quo found in the new organization requires immediate action. Having recognized the risk, potential damage and the need for strategic direction, it would not be enough for senior management to initiate a BCM program. The extent of the task is more important: how much work would be needed to accomplish the level of BCM that the CEO has seen in other firms? What is the required amount of time and money to reach this level of BCM maturity? By answering the CEO's questions, the audit transcends the traditional framework of stating findings and issuing non-quantitative recommendations. The report on preparatory audit activities is more likely to describe the steps towards achieving a major management goal, and to take a more consulting-minded view on BCM.

The Feasibility Study

Another type of audit activity, the *feasibility study*, elaborates on the concepts of compliance and preparatory audit. As a prerequisite, senior management will be aware of the fact that weaknesses in business continuity exist. One step further, members of the management committee will have formed a fairly comprehensive idea of a business continuity project, perhaps occasioned by a preparatory audit. Ascertaining feasibility of this concept – can it be accomplished given a finite set of skills and resources – requires extensive experience on part of the auditor, as senior managers of the entity under review may rely upon the final recommendation. In sharp contrast to a compliance or preparatory audit scope, the auditor will be required to offer professional judgment on known findings and known recommendations. While it is debatable if the term "auditor" would still apply in this situation, this type of "audit" happens fairly frequently, not least because of the scarcity of experienced specialists. When comparing business continuity with other disciplines, it is perhaps not surprising that an organization wishing to start a BC initiative might turn to their trusted (financial) auditors or management consultants to address the question of whether it can be done, as compared with other companies or precedents.

Following the example of the preparatory audit initiated by the new CEO above, an idea will have been formed on "what needs to be done." The feasibility audit adds to this information by explaining if it can be done with current resources, and "how it should be done." The auditor, taking the position of a BC manager (see the Introduction part) might wish to make recommendations concerning internal and external resources, budgets and even project planning to provide clear guidance to the CEO who must decide whether the time and effort for the proposed BCM project is justified in the light of broader strategic objectives.

Given a due diligence situation or imminent strategic changes in an entity under review, business continuity management and its financial background may be of interest to any number of individuals or groups. The example of an initial public offering illustrates how public perception of business continuity – particularly in high-visibility industries – might influence prices and early quotes. Assuming a different scenario, a pre-merger audit at the behest of the potential buyer might reveal severe shortcomings in business continuity planning, indicating increased restructuring costs and increased risk intrinsic to the proposed merger. In an analogy to the feasibility study above, the question is not whether it can be done, but usually whether it should be done.

Consider the case of an e-Commerce firm seeking an additional financing round with venture capitalists. If, for some reason, disaster recovery for critical systems is weak, and corresponding business continuity measures have not been envisaged, the potential investor may be discouraged, given the disproportionate level of risk and exposure. Conversely, if a large industrial firm is contemplating acquiring an international competitor, the due diligence question may be whether the two very distinct BCM cultures can be merged without creating cultural rifts and antagonism. This may be an important issue when looking at manufacturing operations and the way in which continuity is assured at the factory floor level. For instance, the notions of manufacturing "quality," "safety" and related regulations may well become the decisive issue in the merger of two automobile companies.

At the other end of the audit spectrum, forensic audit activities may become necessary, particularly in the aftermath of a disastrous event. The sole purpose is to gather evidence and secure it in an appropriate manner. Business continuity project documents, plans and even logs of plans invoked and actions taken will be readily available. The term "post mortem" is sometimes used – this without any cynicism – to denote this type of audit. The main concern is to secure as much evidence as possible, to enable others to determine a) whether business continuity management was in place prior to the event and b) whether the management process worked when invoked. While often called as expert witness later, the auditor should beware of drawing conclusions from the facts encountered, unless asked to do so. The almost inevitable legal angle to forensic audits of any nature provides for a multitude of formalities and procedures to be followed, thus requiring both business continuity experience and, in most cases, legal advice.

There have been many examples in recent history where forensic activities centered around business continuity, emergency management and securing related facts and evidence. The events themselves have been extensively researched, for instance the King´s Cross Fire in the United Kingdom, various railway-related disasters in European countries, the Hillsborough and Heysel catastrophes during soccer matches, and many others. It is less well known how business continuity issues – whether crowd control, escape routes, physical damage and casualties, financial losses or public relations questions – were prominent topics in official inquiries and the work of official commissions. However, the forensic perspective on auditing business continuity is omnipresent when it comes to substantiating insurance claims, evaluating legal proceedings and giving evidence.

The scope of business continuity audits should ideally be expressed in one single question, depending on the framework and the auditee's expectations. The following examples indicate how the decisive question might be formulated:

- Does it follow the rules ?? (*Compliance*)
- What needs to be done to bring it up to scratch ?? (*Preparation*)
- Can it be done with what we have ?? (*Feasibility*)
- Is it worth the effort ? Is it worth buying ?? (*Due diligence* etc.)
- What happened ? Did it work ?? What went wrong ??(*Forensic*)

While the examples outlined could be applied to any audit, they clearly show that a scope along the lines of "have a look at business continuity," although often applied, is unlikely to yield the expected result. The scope, once defined and agreed upon by auditor and auditee, should be formalized and documented to avoid any misunderstandings.

Step 2: Determine the question that should be answered by the audit

Historically, there has been an unfortunate tendency to equate business continuity as an overall concept with disaster recovery for information systems and technical facilities. In spite of efforts by governments, professional associations and auditors, this common misperception is very much alive and even dominant in certain countries and cultures. In the context of corporate environments, typical statements pointing to the presence of an IT bias might include, but not be limited to:

What is said	***What it probably means***
"That's the IT manager's responsibility, talk to him / her"	I don't understand what those auditors are talking about, but I'd better be seen doing something. Sounds technical, might be best if I send them over to the IT department.
"Yes, I agree that we must do something about our bookkeeping systems."	Whatever continuity is, the auditors usually look at our financial systems. Right, our ERP has weaknesses, so I'd better admit it.
"Yes, continuity is important. Why don't we start with the IT function"	This is like the stuff we did for Y2K. All right, I know how to deal with that. Shouldn't contradict the auditor, though, should I now? Continuity sounds good anyway.

Even in multinational conglomerates with a firmly grounded business continuity process at headquarters, subsidiaries and entities in comparatively remote locations may still adhere to the belief that continuity is yet another word for IT recovery. For the experienced auditor, this may pose some unexpected and cumbersome difficulties in clarifying audit scope. In the extreme, the standardized audit program presented in this book may not even be applicable, due to the highly specialized and technically complex challenges of an in-depth IT audit[16]. As a consequence, both the auditee's IT personnel and their senior managers may be disappointed by the comparative lack of technical expertise on part of the auditor, or – conversely – be entirely out of their depth when confronted with "real" BC questions that no one has ever heard before, and even less expected.

> Imagine a situation where an organization is operating widespread and complex IT applications, networks and systems. Without the assistance of a highly specialized IT

[16] CobIT, for instance, features more than thirty so-called "control objectives" . Each control objective, e.g. DS5 "Systems security" is a universe of audit items in itself. Conversely, the BCM audit program may mention "IT" as one of ten infrastructure processes within an organization, without giving it any precedence over, say facilities management.

audit force, it is clear that any attempt at making sense of this will fail. The business line managers see themselves as users of the IT infrastructure, and apart from being charged for this use by the finance department, there is no link between business and IT. In the background, a substantial part of the IT budget has been committed to creating high availability, fail-safe systems that will automatically switch to alternative operations if a technical failure occurs. Imagine further that a non-technical disaster such as a major fire happens. Employees are evacuated, assemble at mustering stations, and the headcount reveals that there are no casualties to be dealt with. Meanwhile, the IT environment of the organization in its fireproof cellar is running in normal mode. Gradually, line managers realize that access to their IT applications, or even to their desktop computers, is limited due to the fact that emergency services have declared the building structurally unsafe and condemned. Likewise, no alternative office space is available, and only a few users have notebook computers that they could use at home.

This example illustrates that IT disaster recovery, although important, is just one part of business continuity management and should be treated accordingly. The damage arising from non-availability of IT systems may be high, but the potential damage of not preparing the other parts of the organization for a disaster is much higher. Only where both components have been considered will the organization be able to survive.

For the auditor, this question is vital, both in terms of scoping the audit and the skills required. Reviewing IT disaster recovery in isolation is a narrow task that should be defined as a subset of business continuity. Conversely, auditing business continuity should focus on the whole business and drill down into the IT function where appropriate.

Step 3: Make sure that the audit is not perceived as an IT-related exercise

Further steps in determining audit scope may be necessary, depending upon the framework and other external influences. However, the overall scope statement should not pre-empt more detailed objectives which may be known before the audit, or arise as a direct result of findings. Stated in broad terms, the audit scope is usually perceived as an agreement between the auditor and auditee management. Within the realm of finance, it may safely be assumed that even the most senior managers possess a reasonable understanding of audit details, not least because of their own experience and skills. Conversely, business continuity management and the related discipline of risk management, often represent uncharted territory to the auditee, thus requiring a more detailed scope statement. Depending upon the level of maturity found in the organization to be reviewed, the auditor may well be required to "sell" the audit of BCM, as it represents one of many "audit projects" that management may choose from. The corporate officer in charge of audit, or the audit committee, is required to perform a set number of reviews each financial year, trying to cover as many open questions and issues as possible. Leaving aside the purely financial (traditional) audit steps mandated by law, other interesting points may emerge, and the audit committee members are certain to conscientiously note these points as a matter of good corporate governance. However, taking into account the limited overall audit budget, it is difficult to determine whether business continuity as an audit objective should take precedence over IT security or factory floor inventories. The scope statement for a BC audit should therefore clearly outline the corporate governance issues addressed, such as determining business risk, examining levels of potential liability, and so forth.

Audit Areas (Modules) and Planning

Business continuity planning and management incorporates elements from multiple disciplines, and the corresponding skill sets required to assess these. Rarely will the auditor be able to personally evaluate and criticize all areas of a complex corporate environment. Even when relying on accepted standards, "connecting the dots" may be very difficult indeed at an in-depth level. Consider for instance the specific knowledge needed to assess either the acceptability of escape and evacuation routes from a building, or the adequacy of post-trauma counseling arrangements made by the human resources department. It is therefore suggested that the audit be subdivided in a modular fashion. One or more of the modules may have to be covered with at a certain level of granularity, allowing the definition of a simple matrix:

Audit Area / Module	Auditor	Expert assistance / 3rd party
Area 1 (BCM Definition)	existence, completeness, adherence to standards, framework, stakeholder management, project management...	possibly: verify the auditor's criteria in overseas subsidiaries possibly: verify the auditor's criteria for auditee's business partners
Area 2 (Risk Analysis)	existence, methodologies used, recent documentation, risk quantification, insurance etc.	possibly: actuarial opinion on insurance policies, external data on risks and events, government or institutional historic data...
Area *n*	...	...

If the auditor is not an integral part of a very large organization, it is unlikely that the variety of data and external skills required will be readily available. The sample matrix clearly shows how networking and subcontracting considerations ought to precede the actual audit. In the homogeneous environment of financial audit or specialized consultancy, this will rarely be an issue, but it is certainly worth mentioning in a field as complex as business continuity management.

Following the ten standardized areas of the BCI / DRII Professional Practices in business continuity, the separation into a matrix of "self vs. 3rd party" is reasonably straightforward[17]:

1. Project Initiation
2. Risk Management and Evaluation
3. Business Impact Assessment
4. Emergency Response and Operations
5. BCM Strategy
6. Detailed Planning
7. Training and Awareness
8. Maintenance and Exercise

[17] A detailed worksheet is provided in the appendix.

9. Public Relations and Communications
10. Coordination with Public Authorities

Each of these areas may be treated as a module in itself, although some interdependencies inevitably exist. The diverse questions arising from the Professional Practices will undoubtedly require an audit approach that involves expert advice and external assistance. However, the mere existence of a common set of modules is advantageous inasmuch as specialized practitioners can be identified and pre-qualified through the standards body itself[18]. Similar to internationally recognized associations concerned with financial[19] audit or IT-related matters[20], the networking aspect to preparing a BCM audit is likely to be a critical success factor.

In a practical sense, the usefulness of third parties or external experts is often questioned: there may be budget constraints, or the auditor may perceive an external resource as a competitor. The commercial aspects of audit are certainly worth considering, although the ultimate goal should be to assemble a well-balanced and sufficiently skilled audit team. In some cases, maximizing audit revenue, or instrumentalizing the project to insert relatively unskilled junior audit staff, have been put forward as reasons for not using an appropriate (perhaps expensive) external consultant. While "training on the job" is an acceptable practice, the distribution of tasks and work packages will show if and to what extent the internal skills of the audit team can satisfy the requirements. Without an adequate and well-rounded set of skills and experience, the auditor may run the risk of having their professionalism questioned by the auditee organization.

Step 1: Determine audit areas and requirements for expert / 3rd party assistance.

Having established the audit areas and the degree of involvement of third parties, the resulting matrix – in conjunction with audit framework and scope – will determine the granularity of audit steps to be performed. In order to comply with typical constraints presented by the timeframe and budget, the auditor may need to restrict the level of detail for one or more of the areas identified. The question of granularity and the validity of the emphasis placed upon certain aspects of BCM have given rise to criticism[21] and should be treated accordingly. It should be noted that confirmation of the existence of certain BCM components does not necessarily mean that they are functional and credible. However, it is often more difficult to ascertain adequacy than existence and completeness, particularly with business continuity. Plans and management procedures not yet exposed to a real incident or disastrous situation may prove impossible to audit in terms of typical notions such as "appropriate," "sufficient" or "adequate." Experience dictates that continuity management, in the absence of at least one instance of "real" use, can only be audited against theoretical standards. While this fact must not be held against the auditor, the issue of granularity – superficial audit with limited budget, versus extremely detailed audit on a point-in-time basis – will undoubtedly arise if and when an event occurs.

[18] For details concerning the BCI / DRII approach and other proposed standards refer to the bibliography.
[19] AICPA and other national / international bodies.
[20] e. g. ISACA
[21] The main lines of criticism appear to follow two distinct patterns. Firstly, the "tick box" approach (cf. Continuity Magazine, June 2001) is said to provide false assurance as to BCM validity. Secondly, the existence of "pre-fabricated" or "one-size-fits-all" packages and solutions to BCM – for instance, standard software-based documentation – is said to be a mere smokescreen to fend off the auditor (LEVITT (1997) among others). Both criticisms, regardless of their validity in individual cases, should be borne in mind when "setting the scene" for a BCM audit.

Once again, the commercial background of external (and most internal) audit projects should be considered when defining the specific objectives and deliverables. Undoubtedly, there will be trade-offs between what to offer the client (internal or external) and the corresponding effort in terms of time and resources. In order to "get the job," there may be a very real temptation to plan for a massive BCM review to be carried out within a limited number of person-days. Several instruments for justifying the plausibility of such "loaded" audit programs are available: the auditor may calculate on the basis of an eight-hour day (whilst planning to work twelve hours per day), the daily rates may be "adjusted," and the collection rate for actual hours worked may be set at less than 100 per cent. However, the time needed to perform specific audit tasks will not change in line with commercial parameters: there is limited room for maneuver, and any audit activities that have been underestimated will cause time and budget overrun when the practical audit takes place. It may be possible to achieve a higher throughput or level of efficiency by enlisting the services of very experienced auditors who may even be familiar with the target organization, but the trade-off between audit depth (granularity) and corresponding effort still holds.

Granularity, and the ultimate decision on the level of detail for BCM areas, should therefore be the lynchpin of a business continuity audit. Contrary to the highly demanding objectives set by standards commonly used, some limitations will almost always apply. The hallmark of the auditor's professionalism is to concentrate the limited set of financial resources on the most interesting, high-priority audit items, as agreed by management. Instead of covering as much of an implausibly broad audit program as possible in the limited timeframe envisaged, the objective should be to select the right areas for "drilling down" and producing meaningful results. High priority items may merit closer examination, whereas some (mutually agreed) areas may not require more than a high level review.

Step 2: Define granularity / level of detail required, as mandated by audit framework, scope and maturity. Document this and obtain senior management, Board or Audit Committee approval.

Thus equipped with a detailed audit plan per BCM area, timeframes, budgets and effort estimating may be done in a realistic manner[22]. Notwithstanding any overarching constraints, such as senior management directives, the audit itself can now be mapped and qualified in terms of time and cost involved. The table below illustrates an extended matrix, which may serve as an outline for estimating[23]:

Area / Module	Detailed Step	Auditor's Effort	3rd Party Effort	Auditee Effort
Project Initiation	PM Method used	2 hours	nil	2 hours
	BCM Project Charter	1 hour	nil	2 hours
	BCM WBS	1 hour	nil	1 hour

[22] For estimating techniques, cf. Kerzner (1998) and other standard works on project management.
[23] For sample worksheets please refer to the appendix.

The detailed matrix for estimating should reflect the issues and questions of the audit program used (*see Part II*). The above example shows that internal (auditee) effort is not to be neglected. While the auditor may not be limited in spending time and effort on individual steps, the auditee "counterpart" will often be much more constrained, due to the requirements of day-to-day operations. Actual effort is usually determined by a number of factors, such as previous audit experience on the part of the auditee, maturity level, organizational structure, etc.

As mentioned before, the "realism" of effort estimates is dictated by the impartial application of professional methods that have been used for several decades. If the process of compiling the matrix is influenced by ulterior motives, such as competitive budgeting and bidding against other auditors, the results at the time of conducting the practical audit may be seriously damaging to the reputation of the team and the auditor's organization as a whole. Although it is recognized that there must be a balance between high and low points of an effort estimate, the resulting timeframe and budget should not be driven by commercial motives only. The scope statement (see above) should have outlined a solid basis for detailed project planning: if the commercial framework has been set down from the very beginning, senior management will know what to expect and what to tolerate in terms of internal and external effort: "you get what you pay for."

Step 3: Define timeframe and budget.

Business continuity is a discipline with which the auditee staff may not be familiar. In many cases, a general level of awareness and understanding will have to be created by means of kick-off meetings or preparatory sessions. In terms of budget and timeframe, these introductory steps may increase effort and resource requirements. In many cases, the preliminary phase of an audit will be lengthy. Auditors, management and a limited circle of technical specialists will discuss the scope and planning issues surrounding the audit. It is very likely that – particularly when abbreviating words and whole phrases after a while – some type of specific jargon, well-known acronyms and unconscious references to certain parts of the documentation will emerge. When deploying the audit, this may be difficult to understand for divisional and departmental staff who have not been involved in the preliminary activities. For instance, a verbal or written reference to the "Area 3 plan for Austria," as seen from the point of view of the line manager, may sound vaguely threatening rather than reassuring. A workshop outlining the areas and the country audit plans would have resolved this problem by openly addressing what the audit is and what the areas are, but the time and effort required to prepare and conduct such a workshop must be taken into account when estimating the overall timeframe and budget.

Example of Audit Framework, Scope and Planning Statement

The following pages present a sample audit plan containing framework description, audit scope and modular planning. While the sample plan may be seen as representative, other formats and modifications may be appropriate, depending upon the circumstances and auditee environment. Such modifications may occur in terms of detail, page layout, wording or other items.

Comments and explanations are set in italics.

Electronic Widgets Plc
Business Continuity Management
Audit Plan

1. Audit Framework

The Audit Committee and the Board of Electronic Widgets Plc have agreed to conduct a review of business continuity management (BCM) covering all divisions and affiliated corporate entities. The organizational units to be audited are subject to a general audit review following the **CobiT** framework (for the IT perspective), **BS 7799** (for the security perspective) and an internal model of **Operational Risk Management** (ORM).

[This paragraph contains project sponsors, explains senior management commitment and defines standards to be followed. It is an essential part of managerial communications to all employees and therefore as brief and simple as possible. A more detailed statement of sponsorship and management support may be needed in some cases, as part of a project charter].

All BCM audit activities should satisfy the criteria set by the aforementioned standards and frameworks, with a view to optional certification. The results of the BCM review shall form part of the global audit findings and the corresponding recommendations. Responsibility for the BCM review has been assigned to a joint project office consisting of Internal Audit (IA) and our external audit firm.

[The framework is defined and explained: BCM findings are part of overall audit findings, and project managers have been appointed. Auditee staff now have a link to their Internal Audit department for questions and know that external audit personnel may be involved].

Auditing BCM, within the larger framework of auditing all business activities of Electronic Widgets Plc, serves as a milestone towards the planned merger with German Widget Factory AG, where similar reviews shall take place. Adherence to the agreed BCM audit program is therefore an essential part of corporate strategy.

[The strategic objective is clear and unambiguous. Employees know that the other party to the proposed merger is conducting a similar review. The importance of the audit is emphasized].

In order to establish transparency and to avoid misunderstandings, an Audit Intranet has been created. This network contains further information relating to standards, the planned audit program and other relevant data. All employees at Electronic Widgets Plc are invited to use the Audit Intranet as a continuous source of information, and to actively contribute to keeping it up to date.

2. Audit Scope

Business continuity represents an important strategic objective for Electronic Widgets Plc. It ensures that all corporate divisions and their business functions will continue to operate under adverse conditions such as major incidents, disasters or disruptions. This may be accomplished by means of preventive measures and a swift, decisive reaction to a critical event.

[Shorthand definition of BCM as a basis for defining the scope.]

A set of ten areas of BCM has been identified to describe the scope of the review, in line with the Professional Practices document published by the Business Continuity Institute and the Disaster Recovery Institute International:

- Project Initiation and Management
- Risk Analysis and Evaluation
- Business Impact Analysis

- Developing BCM Strategies
- Emergency Response and Operations
- Developing Detailed BC Plans
- BCM Training and Awareness
- Exercises and Maintenance
- Public Relations and Communications
- Coordination with Public Authorities

The BCM review shall cover these ten items and report findings accordingly, for the following areas where Electronic Widgets Plc is present:

- Asia Pacific Region
- EMEA Region
- Eastern Europe (Czech Republic, Russia, Ukraine)
- United States

[This is a sample list of regions and countries. Under normal circumstances, an organization chart or a more comprehensive list of corporate entities should be used. As senior management will undoubtedly possess such an overview chart, it should be incorporated into the audit scope statement to eliminate any doubt as to the local implications of the review].

Given the Professional Practices criteria, CobiT and BS 7799, different levels of BCM maturity have been determined by local management. These will be indicative of the depth and level of detail that individual reviews must accomplish. The audit programs agreed locally should take into account the level of maturity reported by local management to avoid inefficiency. These levels are:

- **N/A**: business continuity has neither been discussed nor addressed in a managerial sense. No documentation exists.
- **Novice**: the entity under review is vaguely aware of business continuity, perhaps indirectly in the sense of IT disaster recovery. No attempts have been made to categorize business processes or extend the paradigm of IT recovery to core business activities.
- **Intermediate**: technical recovery has been initiated and partially implemented. Core business processes are not covered by the methodology applied. Management uneasiness is evident, but no steps have been taken.
- **Advanced**: business continuity has been addressed in one or more projects. Some plans exist, and they may be comprehensive. Business continuity, as a notion, is integrated by means of a part-time or full-time position within the organization.
- **Mature**: business continuity management exists as an ongoing process. Plans, concepts and documentation are well developed and up to date. Insular weaknesses in BC may exist.

[Many countries outside the "home" market are likely to be less advanced in terms of BCM. The maturity classification, similarly to the capability maturity model in software development, is a simple device for pre-qualifying audit targets. More specifically, there is no point in conducting a three week audit in a country that has no BCM capability whatsoever. In such cases, it would be much more beneficial to look for the ability to initiate and run a BCM project, rather than "go through the motions". Time savings thus materialized are better spent on auditing countries such as Japan (in our example) who might benefit from having their plans scrutinized and from discussing an already advanced level of BCM with an external expert.]

Region / Country	Maturity Level
Asia Pacific	Intermediate Japan: Advanced; Thailand: N/A; China: N/A
EMEA Region	Novice UK: Mature; Italy: Intermediate
E Europe	N/A
United States	Advanced

[The paragraphs above describe the general areas to be addressed. An indication of maturity for each location / country is given. This should enable the auditor to formulate the detailed program accordingly.]

Electronic Data Widgets Plc is undergoing a due diligence process. The BCM review should therefore address the abovementioned issues in a fair and impartial manner. The major objective of all audit activities is to determine **similarities** and **areas of divergence** between Electronic Widgets Plc and German Widgets AG. This expressly includes an opinion on steps to be taken and required investments to accomplish a common level of BCM maturity for all corporate entities involved in the proposed merger. The results of BCM audits in both companies will be used as a basis for formulating a common BCM strategy, should the merger be concluded. Hence, auditors should bear in mind that different levels of maturity, or even a lack of compliance, may not necessarily constitute an audit finding but should nevertheless be reported accurately and impartially.

[Within the audit framework and scope, the strategic objective is to assist in the due diligence. This means that the audit does not address compliance only. The auditor is to find out whether business continuity management in both companies (in this case Electronic Widgets Plc) can be reconciled to form a common basis or standard. Even if both companies have not yet reached a high level of maturity, and show distinct weaknesses, their "common ground" may nevertheless be evident, thus leading to a favorable audit opinion with regard to the merger. Conversely, if one of the companies to merge recognizes a substantial investment required to bring the other company "up to scratch," this may influence negotiations. (NB:This is one of many examples showing how a due diligence audit scope might be worded. Depending on the situation, it is highly advisable to have senior management formulate, approve and issue such a scope statement!)].

Business continuity management includes, but is not limited to, IT disaster recovery. Steps and plans taken to accomplish a given level of IT disaster tolerance are a **subset** of business continuity management. The audit must therefore not be perceived as an IT-related audit. All business processes should be covered, including core business, support processes and technology-related processes.

[A typical caveat reinforcing the distinction between business continuity and IT disaster recovery. More often than not, it will be necessary to actually spell out this very important fact, particularly if the audit is to cover regions where BCM is not a common notion. However, if local management receives clear guidance from senior (corporate) management or the Audit Committee, it is likely that the audit itself will introduce an understanding of BCM].

Electronic Widgets Plc has recognized the need for up-to-date, expert knowledge in the field of business continuity management. For some of our more complex business processes, the project team will utilize external (3rd party) assistance. External audit steps may involve professional associations, institutions, local government, police forces and other appropriate parties who will actively support the review by contributing their specialist knowledge. The following sections provide a mapping of such 3rd party involvement that should also serve as an orientation for local management.

3. Regional Audit Plan

The following tables give an indication of audit steps to be performed, depending upon the predetermined maturity level of the individual corporate entity. It has been compiled using preliminary information gathered from local management. While most of these self assessments should be considered reliable and accurate, there may be countries where a somewhat different situation persists.

Region	Country / Unit	Audit Areas (Prof. Practices)
Asia Pacific	China	**Area 1** assume that no dedicated BCM is present; collect fragmented plans as appropriate.
	Japan	**All Areas** the audit assumes that a well-rounded set of plans is in evidence. Focus on the BCM life cycle implementation. Document "best practices" that may be used at corporate level to disseminate BCM knowledge after the merger.
	Thailand	**Area 1** assume that no dedicated BCM is present; collect fragmented plans as appropriate.
EMEA	Austria	**All Areas** Planning is likely to be restricted to IT disaster recovery. Verify the IT DR side and look for fragmented plans in core business areas. While Austria may not be advanced in terms of BCM, its strategic importance within the due diligence process is high. Hence, BCM capability must be examined thoroughly.
	France	**Areas 1, 2, 3** Preliminary reporting indicates that IT disaster recovery has been developed and implemented. Other business processes may not be covered by the methodology. Focus on the business impact analysis (BIA) area to determine the extent of BCM awareness for processes other than IT. Calculate buffer time to expand the audit to other areas as appropriate.
	Germany	**Areas 1, 2, 3** Preliminary reporting indicates that IT disaster recovery has been developed and implemented. Other business processes may not be covered by the methodology. Focus on the business impact analysis (BIA) area to determine the extent of BCM awareness for processes other than IT. Calculate buffer time to expand the audit to other areas as appropriate. (NB: in Germany, "early warning" and recovery mechanisms are mandated by the Business Control and Transparency Act 1998. Cover areas of operational risk management to ensure compliance with the Act).
	Italy	**Areas 1, 2, 3** IT disaster recovery has been developed and implemented. Other business processes may not be covered by the methodology. Focus on the business impact analysis (BIA) area to determine the extent of BCM awareness for processes other than IT. Due to Italian legislation in physical and facilities security, a detailed set of technical recovery plans should exist which may be used as a basis for further audit steps.
	Nether-lands	**Areas 1, 2, 3** as above
	United Kingdom	**All Areas** A preliminary self assessment of UK BCM shows a mature BCM lifecycle. Conduct an in-depth audit of all areas and focus on the implementation of training, testing and continuous improvement. Document "best practices" which may be used as generic material at corporate level. Verify compliance with relevant legislation, regulations and industry standards.
E Europe	Czech Republic	**All areas** Preliminary information indicates a detailed level of Y2K continuity planning whose elements have been used as a basis for extended BCM.

Region	Country / Unit	Audit Areas (Prof. Practices)
	Russia	**All areas** Preliminary information indicates a detailed level of Y2K continuity planning whose elements have been used as a basis for extended BCM.
	Slovakia	**All areas** Preliminary information indicates a detailed level of Y2K continuity planning whose elements have been used as a basis for extended BCM.
	Ukraine	**Area 1** Business continuity has not been addressed in a formal manner. Technical recovery activities to date have been limited. Some incidents and major disruptions have occurred over the last six months.
United States	W Region	**All Areas**
	Central Region	**All Areas**
	E Region	**All Areas**

[This table assumes that a preliminary survey has been carried out at senior management level. In cases like this, it is likely that international managers would address the question in monthly or quarterly meetings. However, as indicated by the wording, this may not always give an accurate or complete picture. The "rough" audit plan allows some time and effort estimates and highlights countries and / or regions where additional questions or clarification is required prior to the audit.]

The following table presents a list of corporate and local contacts, both internal and external. Depending on the local situation, it may be necessary to identify and use additional third party resources.

Country	Internal Contacts / Affiliated Entities	External Contacts
China	M Chang, BCM Coordinator Bloggs & Bloggs, Corporate Solicitors	Yu Consulting, Security and BCM consultants
Japan	Mr Tanaka, BCM Coordinator Bloggs & Bloggs, Corporate Solicitors	
Thailand	...	...
Austria	...	...
...	...	...

[This sample table would normally be completed in detail by corporate BCM staff and senior management. It is an essential part of the overall scope and planning documentation, both in terms of responsibility and budget. Contacts not on the list may a) not be authorized contacts (particularly external firms) or b) may not cooperate at the time of conducting the audit. Nominating contacts and authorized third parties is an integral part of senior management sponsorship for the project.]

Example of Individual Audit Program

The above example summarizes the preliminary statements and steps initiated by senior management, the audit committee and the auditors themselves. Given this information, individual audit programs can be set up for each country on the list. Considering the situation suggested by the sample statement, it may not be advisable to have senior management or a corporate audit team define local audit programs[24]. As a next step, the auditor would most likely seek advice and discuss details with the contacts nominated for each country. Draft audit programs would then be presented to senior management at corporate level to define and approve an overall program with timeframes and budgets.

The following example shows how planning might be effected for one of the countries on the list. It should be noted that, at this stage of audit planning, the main objective is to establish a project plan rather than to determine the **contents** of each audit[25]. While the sample program below is in English, the auditor might consider translating it into the relevant local language(s) to reduce the potential for error or misunderstandings. At first sight, such an exercise may appear cumbersome. However, the language "issue" will produce interesting discussions on wording, document structure and individual audit steps, which may ultimately help in bridging any cultural gaps[26].

Electronic Widgets Ges. mbH
Vienna, Austria
Business Continuity Audit Program

Following the corporate audit plan for business continuity management (BCM) for the Electronic Widgets Group of companies, Austria forms an important part of the EMEA region. The audit should therefore encompass all areas of business continuity as listed in the BCI / DRII Professional Practices. Austrian legislation, regulations and industry standards will be included as appropriate.

An audit team has been formed under the leadership of the Austrian BCM coordinator, assisted by corporate internal audit and external auditors. The organization chart below outlines the team structure.

[24] As with any other corporate initiative, empirical research indicates that "buy-in" at local levels is commensurate with participation and representation of local interests. In the situation set by the example, the cultural and regional diversity is significant, which strongly suggests a localized approach towards the audit.
[25] For a detailed discussion of contents, refer to the model audit program in section II.
[26] Cultural and linguistic diversity have been treated extensively in management literature, and the reader is referred to standard works for further data. English in particular, despite its global use as a "working language," often acquires a localized idiomatic nature (U.K. vs. U.S., West Indies vs. India and so forth) which impedes technical discussions and project management.

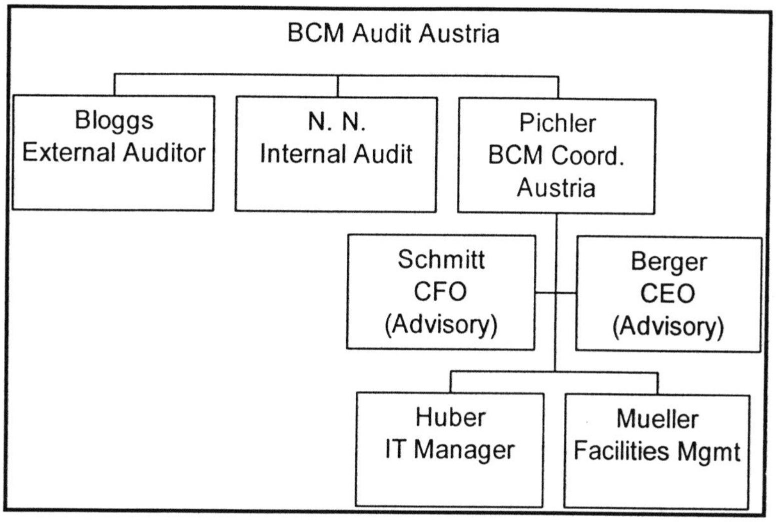

Due to the technical complexity inherent to business continuity management, a list of requirements for external participants has been agreed. These third party service providers have been selected on the basis of proven expertise in BCM, and service level agreements have been set up by the Austrian BCM team. The table below describes roles and responsibilities for each area of BCM audit.

Audit Area / Module	Auditor	Expert Assistance / 3rd party
Area 1 (BCM Definition)	all activities	none
Area 2 (Risk Analysis)	project leadership, data gathering, evaluation of local risk analysis and management activities	1. Insurance companies (coordinated by Austrian insurance dept.) 2. Meteorological Institute, Vienna University 3. Wienstrom, Wiengas (local utilities, coordinated by Austrian facilities mgmt. dept.)
Area 3 (Business Impact Assessment)	verification of Austrian BIA activities	suppliers and service providers to Electronic Widgets Austria, coordinated by Austrian purchasing dept.
Area 4 (Define BCM Strategy)	all activities	none
Area 5 (Emergency Response and Operations)	project leadership, data gathering, evaluation of local concepts and plans	1. Vienna Police 2. Fire Brigade 3. Local Government all, coordinated by Austrian BCM coordinator
Area 6 (Developing Detailed BCM Plans)	project leadership	1. SCM – local BCM consultants, assistance with software-based planning tool. Reports to Austrian BCM coordinator
Area 7 (Training and Awareness)	all activities	none
Area 8 (Exercise and Maintenance)	all activities	none
Area 9 (Public Relations and Communications)	data gathering, evaluation of local concepts and plans	1. comVienna – local PR consultants approved by corporate management 2. Corporate PR agency, coordinated by external auditors
Area 10 (Public Authorities)	see area 5	see area 5

A time and effort estimate has been prepared on the basis of these assignments. Given the strategic importance of the Austrian company for Electronic Widgets Plc, it has been agreed to cover all areas rather than restricting the audit to areas 1 – 3 as suggested by the maturity level stated.

BCM Audit Program Austria
Project Timeframe and Effort Estimate

Area / Module	Detailed Steps	Auditor (man days)	External (man days)	Internal Effort (Austria)
1	full audit program (section 2)	5	Nil	5
2	full audit program (section 2)	8	5	8
3	full audit program (section 2) including model BIA calculations	10	3	10
4	-- " --	5	Nil	3
5	-- " --	5	3	5
6	-- " --	10	10	5
7	-- " --	2	Nil	1
8	-- " --	2	Nil	1
9	-- " --	1	1	1
10	-- " --	1	1	1
Total		**49**	**23**	**40**

[In this example, Austria has been selected for reasons of brevity. The above listing of time and effort estimates can be expanded to cover more detailed audit steps. For other countries in the sample list, the auditor would be expected to provide a more detailed description of activities (internal as well as external) and the amount of days estimated. The term "full audit program" denotes all questions and audit items in section 2.]

Area / Module	External Effort (ATS / EUR)
1	Nil
2	100,000 /
3	60,000 /
4	Nil
5	60,000 /
6	200,000 /
7	Nil
8	20,000 /
9	20,000 /
10	20,000 /
Total	**480,000 /**

On the basis of these budgetary figures, the level of detail for the Austrian BCM audit should cover all subsections of the standardized audit program. Individual project plans – preferably using MS Project – have been prepared and submitted with the overall BCM audit plan, in order to establish clear milestones for assessing and measuring performance. The primary milestones are listed below, including projected dates and deliverables.

Milestone	Deliverables	Due Date	Responsible
BCM Definition	Assessment of BCM project definition, interim report	end Oct 2001	Bloggs
Risk Analysis	Full documentation delivered, audit report	mid Nov 2001	Pichler, Bloggs
Internal BIA	Complete documentation of internal BIA material	end Oct 2001	Pichler, Bloggs
External BIA	Suppliers, service providers and other third parties: BIA documentation complete, interim report for BIA complete	end Nov 2001	Pichler, Bloggs, suppliers´ representatives
BCM Strategy	Assessment of selected strategy and approval process, interim report	end Nov 2001	Pichler, CEO, CFO, Bloggs
Emergency Response, Coord. with Public Authorities	Documentation, review of processes and in-depth analysis of relationship with public stakeholders, separate report	end Nov 2001	Bloggs, local authorities and other public stakeholders
Detailed Plans I	Assessment of planning process and development controls	end Dec 2001	Bloggs, Pichler
Detailed Plans II	External (SCM) report on software use and status of software-based plans; combined report for detailed planning complete	mid Dec 2001	SCM, Bloggs
Training / Awareness	Review of all training activities and training plans, report on general BCM awareness	end Dec 2001	Bloggs, Pichler, department heads
Exercise Planning	Assess exercise planning process, partial report	end Jan 2002	Bloggs
Exercise Action	Conduct exercise as part of audit, report results	end Jan 2002	Bloggs, Pichler, departments
Maintenance Planning	Assess maintenance process and issue partial report	end Jan 2002	Bloggs
Maintenance Audit	Report on maintenance following the exercise action step; combined exercise and maintenance report completed	mid Feb 2002	Bloggs, Pichler
PR and Comms	com Vienna Report complete Corporate PR report complete	mid Feb 2002	Pichler, external agencies
PR and Comms audit	Assessment of historic events and PR-related issues, combined report (with com Vienna and PR items)	end Feb 2002	Bloggs, Pichler

These milestones are indicative of the timeframes envisioned for each audit deliverable. It should be noted that not all activities will be carried out on a full-time basis, to avoid placing an unnecessary burden on line departments and BCM coordinators. The deliverables are in line with the overall timeline for the due diligence / pre-merger activities.

The sum of all audit plans will provide a useful overview of activities under the overall audit program. While the scope statement is binding on all entities involved, the individual plans define the level of detail required to accomplish the audit objectives. Within the limits of individual (country) plans, the questionnaires and techniques applied are likely to show considerable variance. The sample questions in section II of this book are designed to point out some areas of interest, although the factual assessment and interpretation of answers is still directly linked to the auditor's experience and level of BCM skills.

Summary

- Select, discuss and agree to a **general framework** for the proposed BCM audit. Business objectives "driving" the requirement for a BCM audit should be the decisive factor in selecting the right approach.
- The BCM auditor should be aware of **standards, codes of practice and other established models and frameworks** already on the market. Most of these standardized frameworks offer distinct advantages compared with proprietary solutions that re-invent the wheel.
- Depending upon the situation, the audit framework may be governed or restricted by **legal issues**, or by statutory requirements. The BCM auditor should respect the order of precedence by adapting any audit framework to these constraints.
- Define the audit scope, using a **three step approach**
    - **Step 1**: determine the level of maturity of the entity under review
    - **Step 2**: what is the question to be answered by the audit (compliance, feasibility, evidence etc.)
    - **Step 3**: make sure the audit is not an IT-related exercise. Otherwise, it should be renamed to reflect the true nature of audit objectives and activities
- Identify any external assistance needed to perform the audit
- Define the level of detail / granularity for each audit item. Differentiate geographically or by other categories where appropriate
- Set budgets and timeframes, based upon individual audit plans

- Formulate the **framework and scope statement**. Obtain senior management approval. The framework and scope statement should be issued by senior management of the auditee, not by the auditors themselves.
- Formulate individual **audit plans**, describing areas, depth of analysis and local contacts. Identify areas of interest that have the highest priority for management.

Further readings and typical audit **frameworks** are listed in the Bibliography section.

2

CONDUCTING THE AUDIT

As an auditor, you are managing the practical phase of a BCM review. Chapter 2 explains how to schedule the review, how to estimate time and effort, and how to streamline the process of formal audit steps. Known difficulties and pitfalls, many of them unique to BCM, are explained in detail. Even if you are a seasoned audit professional, this chapter may help you in identifying typical problems associated with reviewing a complex process and interacting with a wide range of managerial and technical responders. As a business continuity manager, Chapter 2 may help you understand the challenges presented by reviewing the BCM concepts without actually managing them yourself.

Scheduling and Administration

Having planned the overall audit framework, scope and individual modules – perhaps multiplied by a number of audit targets – the execution phase of the BCM audit requires additional planning. In contrast to the initial setting of constraints and logical boundaries, conducting on site activities is largely based upon previous experience, an individual auditor's perception of the situation encountered, and effort estimates gathered from similar projects. For a less experienced audit team, these considerations may pose unexpected problems. For instance, estimating durations of interviews, administrative overhead or average times allotted to compiling a documentation folder are rarely taken into account when planning for on site presence. As a result, part of the audit work – more often than not – takes place at night or during weekends. There may be other reasons that lead to a "compressed" timeframe: overly ambitious project goals set by the auditor to win a competitive bid, increased pressure by the auditee management or audit committee, and even scope creep introduced in the early stages of the audit project. The end result is almost always identical – audit quality is compromised due to unforeseen pressure and the need to meet tough deadlines.

Despite the fact that the timeframe and budget should have been finalized at this stage, the scheduling and project administration phase of the audit are often questioned both by auditee management and project managers within the auditor's company. In an attempt to maximize productive hours, and to generate follow-on projects from audit findings, the hours spent on planning and documenting are sometimes regarded as unproductive. However, once the overall budget has been set, saving time and money by cutting corners is likely to lower the level of audit excellence rather than achieving it through being "earlier than on time" and "cheaper than promised!"

These basic truths, although applying to almost any project plan and its subsequent execution, acquire a particular significance when it comes to BCM, which extensively uses and relies upon human interaction rather than the study of documented materials. A direct comparison with financial audits would undoubtedly reveal a somewhat "fuzzy" approach taken by the BCM auditor, whereas the financial auditor is often able to resort to more formal methods of analysis[27]. Business continuity management does not readily lend itself to being analyzed by way of "ticking boxes". On the contrary, even detailed statements made during fact-finding interviews will usually require independent confirmation which may be effected through physical inspection or other suitable tests. The validity of representations made by the auditee organization may not be as clear-cut as a set of financial accounts, and the notion of accuracy is replaced with the much wider idea of adequacy.

As a rule of thumb, the information gathering part of the BCM audit may be subdivided as follows:

- Interviews and requests for information – 50 %
- Filing, perusal – 10 %
- Analysis and interpretation of information obtained – 40 %

If the overall audit objective – as stated in the scope document – is to verify compliance with standards or regulations, the analysis phase may take longer, while forensic activities might require an emphasis on obtaining data rather than interpreting the material. The above listing should nevertheless be indicative of the additional effort for administrative and analytical purposes.

In contrast to financial or other audits, additional means of documentation may prove useful in ascertaining facts about BCM. Photography, video, or scanning of handwritten transcripts (e.g. for exercise results) are among the tools used to record detailed information[28]. However, there is a significant time and effort component involved which should be recognized by the BCM auditor in planning on site activities.

An additional overhead for administrative purposes should be included as a buffer or safety margin, particularly when conducting international audits. Unexpected delays may occur as a result of

[27] in most cases, this is simply the result of a more formalized set of documentation. Both auditor and auditee know what to expect in terms of year-end activities.

[28] Practical experience has shown that photography, including development and filing of negatives (or digital processing) may take considerable time.

numerous factors, irrespective of the budget and timeframe agreed. Practical experience has shown that the auditee entities may be under pressure to perform day-to-day tasks, and senior management intervention often serves as a means to reprioritize or defer business continuity activities. The auditor should anticipate such distracting events and avoid situations which might set conflicting objectives for the auditee organization. Likewise, individual preparation should preempt any difficulties or delays that are due to the general environment in which the audit takes place:

- Issues arising from local requirements, such as visa, accommodation etc.
- Familiarization, e.g. weakness, fatigue, illness
- Cultural difficulties, e.g. negotiation style, monetary incentives
- etc.

While the above may be familiar to international travelers and auditors, it is nevertheless important to note that delays and budget overruns may occur if no provision has been made. In an international setting, translation difficulties and misunderstandings regarding the presentation of results may add to the degree of uncertainty to be anticipated by the auditor.

A typical example is the case of the US audit firm contracting for an international BCM audit covering Europe. While scoping, planning and budgeting may appear straightforward in a national setting, the subsequent deployment of audit plans is often more difficult than anticipated. When sending memoranda to a number of country subsidiaries, the answer to the requests and timescales is "yes." This may imply a number of different shades of agreement, depending on the country involved and the question asked. Administratively, the auditor – when arriving on site – may be presented with a neat office stacked with hundreds of folders and told that this is "the documentation," albeit in a foreign language such as German. Conversely, the negotiating style of another country may be such that the aforesaid "yes" includes no more than wishing to please corporate HQ without necessarily understanding the nature of the audit that has been announced. The times calculated for traveling between locations in certain countries may differ from what has been envisioned during the planning stage, as may the means of transport. In this example involving the US auditor, traveling by train may seem unusual in the US, whereas it is a normal practice in most European countries. Flights, while still common, require elaborate logistics to and from the airports, as the auditee organization may not be situated within the immediate vicinity of a major city. The list of issues surrounding the administrative part of an audit could be continued, but suffice it to say that the original plan is rarely accurate until it has been validated by a person or group with "on the ground" experience.

Example of Interview Schedule and Administration

The following example continues the scenario outlined in the previous chapter. Given that scope and framework lead to individual audit planning, this step determines how on site interviews might take place within the timeframe and budget agreed beforehand. While many auditors prefer tabular (spreadsheet) representations of this planning step, project planning software may prove more versatile in terms of scheduling detailed activities and matching them against budgetary and time constraints.

BCM Audit Program Austria

Interview and Administration Schedule

Topic	Date and Time	Interviewee(s)
Area 1 – BCM Definition and project management	Oct 4, 0900 – 1100	Schmitt (CFO), Berger (CEO) part time, Pichler (BCM Coord.)
Area 2 – Risk Analysis	Oct 4, 1300 – 1500	Pichler, Mueller III (Corp. risk mgmt), Mueller II (Insurance dept.)
Area 2 – Risk Analysis	Oct 8, 0830 – 1000	Wiengas (contact to be nominated)
Area 2 – Risk Analysis	Oct 8, 1100 – 1200	Met. Institute Vienna (contact to be nominated)
Area 3 – BIA	Oct 5, 0900 – 1100	Pichler, Internal Audit, department managers to be nominated, Mueller (Facilities Mgmt)
Area 3 – BIA	Oct 7, 0900 – 1100	Software provider*
Area 3 – BIA	Oct 7, 1100 – 1300	Hardware shop*
Area 3 – BIA	Oct 7, 1400 – 1600	Outsourcing partner*
Area 4 – Strategy	Oct 11, 0900 – 1000	Berger (part time), Schmitt, Pichler, Huber (IT)
Area 4 – Strategy	Oct 11, 1100 – 1200	Systems provider, Huber, Mueller III – backup data center
Area 5 – Emergency Response	Oct 11, 1500 – 1700	Pichler
Area 5 – Emergency Response	Oct 12, 0800 – 0900	Pichler, police representative
Area 5 – Emergency Response	Oct 12, 0930 – 1100	Pichler, fire brig. representative
Area 6 – Detailed Planning	Oct 12, 1300 – 1500	Pichler (part time for introductions), SCM (local BCM consultant)
Area 6 – Detailed Planning	Oct 13, 0900 – 1300	SCM – group session with consulting team
Area 6 – Detailed Planning	Oct 13, 1500 – 1800	SCM – intro to software
Area 6 – Detailed Planning	Oct 14, 0900 – 1300	SCM – detailed software review
Interim status meeting	Oct 14, 1500 – 1600	Pichler, meeting to exchange results and schedule additional interviews if necessary
Area 7 – Training and Awareness	Oct 15, 0900 – 1000	Krantz (PR dept.) – awareness program
Area 7 – Training and Awareness	Oct 15, 1100 – 1230	Schlumberger (HR) – training planning
Area 7 – Training and Awareness	Oct 15, 1500 – 1600	Schlumberger – review recent training schedule
Area 8 – Exercise and Maintenance	Oct 18, ...	...
...	...	...
All areas	Oct 19, 0900 – 1000	Photography
Status meeting	Oct 19, 1100 – 1300	Pichler – meeting to ascertain audit status and determine further action
Documentation review	Oct 19, 1300 – 1500	Pichler – meeting to review documentation obtained and issue missing documents list

The asterisk (*) mark denotes interviews conducted at Electronic Widgets Austria Ges. mbH.

[The interview schedule outlines direct contact with the auditee. Additional administrative steps may have to be taken. Note that the times and duration of interviews are usually scattered across the week. The above example represents an unusually seamless suite of interviews relating to audit areas. In most cases, the contact points with the auditee may have to be spread over a longer period of time (vacation, sick leave etc.). Where the above example states that groups should be present, it may take longer to find a suitable time and date. External advisors and consultants are usually more accessible, due to their being invited by their client to assist with audit activities. According to the 50:10:40 rule described above, the frequency and duration of interviews should be matched by corresponding documentation and analysis activities.
Note that interview duration is limited to a maximum of two hours. Only in exceptional cases will the interviewees be requested to make themselves available twice a day, considering the physical limitations imposed by an interview situation. As a rule of thumb, two hours will exhaust an interviewee who will often face unfamiliar ground. The situation may be different with PICHLER (in the above example) whose main function implies extensive experience and therefore less mental strain when interviewed. Experience has shown that individuals who "know what they're doing" are less likely to experience physical strain as a result of prolonged interviews. It is rare to find more than one or two persons who show this characteristic.]

Topic	Date	Administrative Action / Person Responsible
Flights	Sep 15	John Doe Travel, Inc. (JDT)
Visa (for non-EU/non-US audit staff)	Sep 01	Human Resources
Hotel Arrangements	Sep 20	JDT
Documents Translations	Sep 10	SCM (Vienna)
Vaccinations (where applicable)	Aug 15	local physicians (check!)
Confirmation of audit planning (by Austria)	Aug 10	Pichler (Vienna)

[These and other administrative issues may have to be resolved several weeks, or even months, before the on site audit. In the Austrian example, visa would not normally be required for European Union or US citizens. However, some specialist resources may find that, due to their nationality, visas are required as a formality. In the US in particular, people who have a permit of residence may not be eligible for visa-free entry into Austria if they still hold a foreign passport. Compulsory vaccination requirements may be in force when traveling from or through certain countries. In this example, the Austrian environment would not necessitate extensive administrative effort to prepare the audit. However, other geographic regions may take up to three months of preliminary administrative steps].

The example of an audit and administration schedule illustrates that in addition to interviews addressing items directly related to BCM, project management tasks – as indicated by "status meetings" and other non-investigative discussions – provide structuring and feedback opportunities for auditor and auditee (see below).

Interview Contents

Style and Technique

Given the general nature of an audit, interview often create an atmosphere of investigation, or even interrogation. As issues of business continuity are almost inevitably linked to notions of "faults," "failures" or "disruptions," an adversarial situation easily sets in. Questions highlighting areas where problems or weaknesses exist will often lead to attempts on the part of the interviewee to disguise or misrepresent the facts. In many countries, expenditures for business continuity management are minimized, and staff responsible may be working under enormous pressure. If such a corporate culture is in evidence, the audit is more than likely to provoke resistance and defensive reactions. Conversely, the auditor may be viewed as an external source of credibility, tempting the interviewee to instrumentalize the – technically inexperienced – auditor as a witness to claim additional funding or project sponsorship.

Considering these two exaggerated positions, interview style and contents should focus on situational aspects rather than "filling in the forms," as is often the case. The standardized audit program described in section II of this book should be reviewed accordingly, prior to conducting on-site interviews. To the novice auditor in particular, the experience of being a prying nuisance sent in by headquarters may be quite frustrating, while the feeling of "being taken seriously" when asked to reinforce local concerns to senior management may produce a contrasting feeling of pride and elation. It is – admittedly – very difficult indeed to appropriately confront such personal experience when one is unable to resort to a rich background of previous audit experience. However, the question of personal motives behind specific interactions during interviews ("what's in it for them / me") often serves as a first indicator that helps in establishing a more impartial way of seeing the situation and the facts. In any event, the auditor should go to extreme lengths to avoid "taking sides" or even adopting the intra-organizational thinking of "us and them" that is often found between technical (even BC) and senior managerial circles of the firm. Likewise, the psychological barriers between "techies" and "suits" often seen in large organizations should not lead the auditor to sympathize with the side most closely reflecting his or her own background. For instance, an auditor with a strong IT background might feel that interviewing an IT disaster recovery manager is somewhat easier than discussing BC issues with a senior manager. While the point of view put forward by the senior manager may appear removed and difficult to understand, there may be good business reasons behind it. "The other side must be heard" is therefore a good rule of thumb to use when interviewing.

Assuming that finance and "traditional" audit represent a comparatively high level of abstraction (expressing real life circumstances in terms of money), the contents of a BCM audit interview are much more tangible. Seldom are facts and resulting consequences translated into profit and loss – to the business continuity manager, the factual findings are likely to cause a much more immediate psychological and emotional concern. Possible disaster, critical events, and major damage to facilities may culminate in human suffering, injuries, or even major casualties. A business continuity manager who has identified deficiencies in evacuation procedures may entertain strong ethical doubts about seeing these in a "business context," to the point of being a driven person. Hence, the intrinsically strong tendency to highlight difficulties and problems is often stronger than would be expected in a financial audit. Open question techniques and a general willingness to listen on part of the auditor are likely to evoke quite detailed responses in many areas, particularly where the auditor is perceived as a person or group of persons that are prepared to reinforce the auditee's concerns at higher management levels, or alleviate the personal burden of the interviewee.

When administering an impartial assessment routine such as the standardized audit program, the initial response will almost inevitably highlight areas of business continuity management that have been neglected, or even consciously ignored. However, the auditor should bear in mind that the audit as a project is not primarily designed to augment existing tensions between specific departments and corporate management. When designing an interview, "loaded" or leading questions may indicate an unwelcome bias, whether it is towards "covering up" or "stirring up trouble." Certain aspects of BCM, particularly in the areas of management support and organizational importance of the function itself, often serve as a magnifier for the auditee's expectations: qualitative judgments may arise where, for instance, the auditee feels that a lack of management attention is not classified as a major problem – the auditor is seen as a willing tool for covering up issues at the behest of senior management. Conversely, a detailed statement of certain findings that are deemed insignificant by management may lead senior managers to believe that the auditor has been instrumentalized to "push a hidden agenda" in favor of a specific department. While the level of detail and the presentation of answers to standardized questions is indeed a reporting issue, even the original wording of audit questions in an interview contents sheet may strongly influence, or even bias[29], the end result. Although this may appear to be self-evident, the use of closed or "trap" questions is often done unwittingly, particularly when interviewing about regulations.

In order to retain an impartial position vis-à-vis the auditee, any interview contents should be clearly and unambiguously linked to sources identified in the audit program. Frequent references to acknowledged and recognized standards or regulations are often the only means of providing detailed evidence as to the factual situation. As remarked above, auditing on the basis of standards represents a comparison with an ideal world by definition. In practical terms, the mere question itself may evoke disbelief, skepticism or outright rejection of the topic at hand: consider the – fairly frequent – reply "nobody does that in our industry". The words "does" and "our" clearly denote some form of custom, time-honored tradition and, subconsciously, the implied criticism directed against the interviewer. If nobody does it that way, why then is it necessary to ask the question at all? The auditor, being clearly outside our industry, is not applying the right standards (our industry standards that may have evolved over the years). Basic analysis of semantics will therefore show how the auditee weighs the questions against their own experience and values[30]. Hence, the open application of extraneous standards and regulations – the auditor should overtly and directly refer to these – changes the frame of reference from "our industry" to a more objective context.

Questionnaires vs. Face-to-Face Interviews

Depending on the timeframe and budget agreed for the audit, the question of pre-fabricated questionnaires or individual interviews will often arise. Business continuity management, as a multidisciplinary field of research, does not readily lend itself to the questionnaire approach. While standardization and uniform questions may be desirable in terms of a general comparison between entities and units under review, the multiple facets of BCM – particularly in technology-biased departments or business functions – are rarely grasped by finite questionnaires which might easily miss the nuances and "between the lines" statements providing additional value to the auditor. In a financial audit framework, benchmarks and numbers to be elicited from a finite universe of accounts are usually well defined. Despite this fact, auditors often face divergent definitions of specific terms and vocabulary, thus having to retrace certain audit steps or having to "normalize" responses to questionnaires. BCM, given its diversity and broad coverage of most business functions, has an even higher potential for errors or misunderstandings. The initial time savings seemingly realized by distributing a standardized questionnaire are inevitably offset by ambiguous or misguided responses that cannot be used for further analysis. Consider for instance the popular definition of "collateral damage" and its implications for the person tasked with completing the questionnaire. Although well meant, the questions are bound to confuse the respondent, and multiple definitions of "damage", "collateral" and their subjective meanings in the context of the department will lead to an unintelligible, and unreliable, response.

[29] Specifically, closed questions implying a duplicated negation, (i.e. not allowing a truly positive response), e.g. the classic "have you finally stopped beating your wife?" or "does senior management believe that cutting the fire protection budget will comply with state regulations?"

[30] for more specific treatments of these phenomena, please refer to the extensive body of literature in organizational psychology.

Face-to-face interviews, on the other hand, will often yield surprising results when directed by an experienced interviewer. The potential for error is – by definition – minimized as a result of interaction and immediate clarification. While the time and effort required to conduct in-depth interviews may appear disproportionately high, the resulting data and information are of a much higher quality, allowing both the auditee and the auditor a margin of interpretation that is absent in a questionnaire setting.

In software-based scenarios, some standard packages[31] allow the design and subsequent deployment of customized questionnaires, including a response processing engine. The software will "emit" questionnaire documents and even customized database screens for the respondent who supplies the appropriate data and returns the completed package. Conceptually, this approach represents a distinct improvement as compared with written questionnaires. However, it should be noted that several prerequisites must be taken into account prior to deploying software-based questionnaires. First, the amount of preparatory activities is significant. For the auditor and the auditee alike, the primary objective is to create a satisfactory level of awareness and understanding for such software-based interviews and information requests. Second, the returned items in themselves require careful scrutiny as to their validity and completeness. Only in very few organizations with a high level of BCM maturity will the software-based questionnaire approach produce any meaningful results. The auditor should beware of products and methodologies suggesting a "shortcut" approach towards information gathering.

Despite these shortcomings, large-scale BCM audits may rule out the face-to-face approach and the individual processing of results. The analytical scope of the audit, as explained below, will often sacrifice information density and precision in favor of standardized patterns applicable to many corporate entities or countries. Hence, the auditor should balance this legitimate concern with the need to obtain information in as detailed a manner as possible. It is advisable to refer to the scope (as discussed above) to determine whether standardized response format or exhaustive information have the highest priority.

The standardized audit program presented in this book lends itself to both approaches. While it may be deployed as a questionnaire, including weightings and measurements, the auditor may decide to derive rough guidelines similar to a medical diagnosis obtained through in-depth questioning of the interviewee. It should be noted that questionnaires, once defined, are difficult to modify for practical and psychological reasons. Practical experience has shown that questionnaire designs, once accomplished, are perceived with a degree of finality, taking on the characteristics of a product. Modifications are awkward, particularly after deployment. In contrast to face-to-face interviewing, respondents usually regard questionnaires as a one-time effort. Subsequent modifications to the set of questions will undoubtedly prompt responses along the lines of "why didn't you ask that question before." In many BCM projects, including audit and review activities, this restriction in terms of clarifying or expanding the set of questions, these expectations on part of the auditee are anticipated by the auditor or questionnaire designer. An unwelcome secondary result of this finality of a questionnaire is that project groups and audit managers will often attribute an exaggerated level of significance to a questionnaire, on account of its one-way characteristic: there is only one chance of finalizing and deploying it. Once the questionnaire has been launched, it cannot be taken back or modified. Psychologically, the original objective of the questionnaire – a mere means to an end – takes second priority. In sharp contrast, it is quite common to communicate with an interviewee following a face-to-face interview. The auditee is unlikely to take exception to a phone call or e-mail requesting clarification of a point made at the time of the interview.

[31] cf. von Rössing and Wiring (2000) for a detailed review of selected standard BCM packages.

Regardless of the audit scope and defined objectives, the auditor should take into account these psychological processes pertaining to questionnaires and "mechanistic" devices for information gathering.

Textual Response vs. Multiple Choice

If questionnaires are used as a means of eliciting information, their structure – multiple choice or open response – is often seen as a direct function of time and effort scheduled for the audit. The multiple choice approach would appear, at first sight, to offer some distinct advantages. Such questionnaires can be prepared fairly quickly, allowing a standardized deployment and comparatively efficient processing or analysis. However, the focus of questions and the granularity of responses is usually limited in terms of answers. Open text questions encouraging verbose responses, on the other hand, are often considered time-consuming and less uniform, thus presenting difficulties in results analysis. Hence, there is a clear trade-off between multiple choice and open textual questionnaires.

In the context of auditing a complex business continuity management process, the obvious conclusion is a strong preference for open textual response designs. However, some areas or subsets of a full BCM audit may nevertheless lend themselves to a multiple choice approach, particularly where quantifiable results are expected. Consider the case of facilities management, where well-established technical standards and regulations are to be taken into account when performing an audit. In this situation, the auditor would probably decide that a multiple choice questionnaire is the most efficient tool for verifying fire protection on site. Conversely, when reviewing the concepts and steps taken to ensure employee well-being in the aftermath of a disaster, a multiple choice questionnaire would almost certainly fail to grasp the nuances intrinsic to this subject.

The proportion of open textual response and multiple choice questionnaires should therefore be weighed carefully against the above-mentioned trade-off. From an audit perspective, the trade-off in terms of time and effort is clearly superseded by the information requirements: whether it is multiple-choice or another questionnaire design, the main objective should be the same as that of a face-to-face interview: elicit as much information as necessary, using the best instruments available.

Referring to the cultural differences outlined earlier, the multiple-choice method is often the preferred means of gathering information in the United States, whereas interviews are seen as the predominant form of obtaining a full picture in most other countries. Practical experience has shown that some respondents there will often insert additional choices, or add verbose comment to the response. While this may be useful to the auditor analyzing the questionnaire, it defeats the purpose of having a simple and quick instrument for information gathering.

Example of Interview Guidelines

Given that audit projects can range from individual efforts involving one auditor to larger audit frameworks, the following example takes a modular approach towards interview contents and the balance between face-to-face interviews and predefined questionnaires. It is indicative of the levels of granularity often encountered when applying the standardized audit program. The high-level audit items usually require in-depth interviews, whereas the more detailed items – possibly governed by technical standardization – are covered by questionnaires.

The example assumes manual design and deployment, for instance using Microsoft Word or any other word processing package. When applying automated questionnaire techniques (as described above, using standard BCM software packages), a similar design might be employed, but logistics would center around the question and response engine offered by the software package.

Electronic Widgets Plc

Interview Contents and Standard Questionnaires

This document is intended to provide an overview of audit contents and the approach towards information gathering. It should be applied to all audit plans and country reviews. The uniform methodology for administering interviews and questionnaires is designed to reflect time and budget constraints imposed by the scope of the audit. The emphasis is therefore on providing guidance rather than standardizing the information gathering phase to an extent that would limit the auditor's freedom of action. It is nevertheless recommended that these guidelines be adhered to in order to allow comparison of results and findings.

Face-to-face interviews have been envisaged for the high-level overview and project management activities within the audit plan. Local BCM coordinators and line management are therefore requested to make themselves available for personal interviews. The information gathered from these sessions will assist the audit team in identifying subsequent detailed work schedules and may serve as an opportunity to identify a series of questionnaires to be distributed to local departments.

The following chart depicts a suggested hierarchy of interviews. Related questionnaires, as indicated in the diagram, may be used to streamline the process of gathering data from larger departments or groups of people. In some countries or corporate subsidiaries, it may be preferable to the auditor and local management to conduct face-to-face interviews instead of distributing questionnaires, particularly when local departments are small enough to allow this approach.

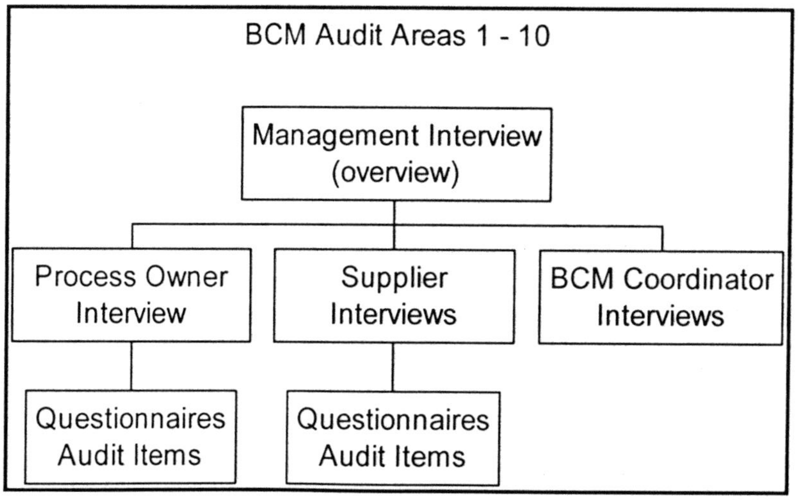

For each interview, a short guideline is included below. Sample questionnaires may be compiled from the standardized audit program, depending upon the audit area of interest. Interview notes should be documented using the template shown below. Questionnaires are subject to consolidation and analysis as described in the "Analysis" document which is part of the auditor's instructions.

Management Interview Guideline

The objective of management interviews is to obtain a high-level understanding of the BCM situation in the entity under review. Management, particularly at senior levels, cannot be expected to have detailed information or knowledge about specific audit items. Rather, the managers interviewed may provide an insight into the overall handling of BCM within the organization, and about the implementation of corporate / centralized BCM programs. Some planning and budgeting information about the BCM process may be available. Specifically, the question of managerial support and BCM sponsorship should be addressed in the course of management interviews. The following list provides a tentative agenda:

1. Audit Presentation
   Outline the purpose, scope and central program for the audit. Indicate the managerial priorities and the objectives. Discuss the audit modules and individual plans as well as timeframes envisaged for on site activities.

2. Liaison / Reporting Requirements
   Agree with interviewees how audit status, completion and preliminary results should be communicated. Local management may wish to have an active involvement and help with overcoming difficulties or obstacles.

3. General Audit Structure
   Discuss further audit steps and their structure. Agree with management how these steps may have to be adapted to the organization under review. Obtain approval for further interviews, including express authorization by management where this is deemed necessary. Present questionnaires and indicate the timeframe envisioned for conducting the information gathering phase.

4. Miscellaneous
   Discuss analysis, reporting and consolidation of audit results. Explain the overall nature of the audit and subsequent steps at corporate level. Ensure that continuing involvement of local management is agreed upon. Note any current or future initiatives for business continuity management which may have been thought about by local management.

BCM Coordinator Interview Guideline

The local BCM coordinator is the focal point of the audit. It is recommended that a series of interviews be held to

- ascertain the status of BCM on a high level
- establish project management and project control
- identify process owners and critical issues
- facilitate organizational awareness of the BCM audit program and its purpose
- clarify open items and agree additional steps

1. Audit Presentation
   Present the audit program and outline the ten areas of the BCI professional practices. Explain the scope of the audit (due diligence) and its implications for the local audit plan. Ensure that there is agreement on the "how and why" of all planned audit steps.

2. Liaison / Reporting Requirements
   Determine a framework for status meetings and communications. Find out if local assistants or BCM personnel can act as "multipliers" to facilitate information gathering and interpretation. Clarify the language issue if necessary. Identify any cultural difficulties and how to reconcile these with the audit objectives.

3. General Audit Structure
   Identify, together with the BCM coordinator, the schedule of further interviews. List important process owners and other contacts for specific audit areas. Map these against the standardized audit program. Make a joint decision on questionnaire vs. face-to-face interviews, depending upon the size and complexity of the local organization.

4. Miscellaneous
   Jointly review the standardized audit program and its planned application to the local environment. If necessary, adjust the original (self-assessment) perception of BCM maturity, existing planning and management processes, and local skills in BCM. Identify any third parties reporting to the BCM coordinator whose roles and responsibilities may impinge upon the audit. Discuss whether external firms or local groups or individuals may have adopted standards or "best practices" that represent a potential for conflict. Discuss legal and regulatory specifics, as well as constraints imposed by environmental factors (e.g. restrictions set by the owner of the premises, government influence etc.).

NB: if there is a local BCM coordinator, it is safe to assume that the person or their department will be knowledgeable and experienced with regard to business continuity. For the auditor, this offers an opportunity to realign priorities and preconceived ideas which may have appeared logical and accurate at corporate level. However, the real situation encountered, particularly in the interviews with the BCM coordinator, should take precedence where appropriate. It is quite possible that, in the absence of knowledge or prior experience at corporate level, new facts may come to light that require some adjustments to the audit plan.

Process Owner Interview Guideline

The business process owners – usually department managers – will usually act as multipliers having to support audit activities at department level. They may be given some instructions by local management and additional guidance by the BCM coordinator. However, it is the process owner who must allocate time and resources to the planned audit activities. Audit steps should therefore be clear, unambiguous, and mutually agreed in order to allow the process owner to realistically estimate overall involvement.

1. Audit Presentation
   Introduce the audit plan for the local entity, and explain the categorization of business processes. Highlight the comparative criticality of the business process being discussed. Explain the "how and why" of individual audit activities. Delineate the contribution the process owner is expected to make to the local audit, and provide comparative illustrations from other audits or companies reviewed.

2. Liaison / Reporting Requirements
   Establish a clear business liaison with the business process owner. Data and information gathering should be explained as an exercise whose results will be shared unreservedly with the process owner, prior to entering the analysis phase. Highlight the need for technical / factual approval by the process owner before questionnaire results are accepted as correct. Discuss further reporting requirements as a function of the process owner's interest: subsequent, consolidated reports including information from other departments may be of vital interest to the person being interviewed. This exercise may also serve to build a bridge between different disciplines in the company under review.

3. General Audit Structure
   Present the set of questionnaires designed for the business process. Discuss the need for uniformity vs. individual process environment with the process owner. There should be agreement on any suggested modifications to questionnaires (to be avoided if possible). Jointly determine the preferred mode of distribution for questionnaires. Define and jointly approve a schedule for sending, processing and collecting responses, including departmental (process-related) deadlines for this exercise.

4. Miscellaneous
   Jointly review the standardized audit program and explain the significance of the ten audit areas to the business process under review. Highlight the areas such as BCM strategy, detailed planning, testing and maintenance where the process owner is required to take an active involvement in all stages of the business continuity management cycle. Clearly distinguish functions or areas such as PR or liaison with public authorities that do not directly concern the process owner. Discuss future actions and budget requirements with the process owner where appropriate, explaining the route from BCM audit towards installing a BCM process in the organization. Ensure that BCM – as understood in the sense of audit as well as ongoing activities – "fits in" with the normal schedule of operations for this business process. Note any concerns and issues verbalized by the process owner to include them in audit reports and further analysis.

Supplier Interview Guideline

Critical suppliers should mirror the BCM capability of the organization under review. It is therefore essential that these suppliers be informed about the business continuity audit and its implications for their business operations, and any existing SLA relationships. The supplier interview serves as a first indicator to assess BCM capability and preparedness. It should be noted that suppliers form an external part of the audit which is not directly linked to the priorities of the overall audit scope and planning. Granularity and depth of these supplier interviews may therefore vary, although the objective of determining BCM maturity remains unchanged. Suppliers, whose contracts with the principal (i.e. the auditee) may stipulate certain performance specifications and penalties, should be seen as serving the interest of the auditee organization. Weaknesses or relevant audit findings should be highlighted and brought to the attention of local management. However, such audit findings need not necessarily be interpreted as weaknesses of the intra-organizational (auditee) business continuity management process.

1. Audit Presentation
   The auditor should present planned activities and salient criteria only inasmuch as they relate to the supplier being interviewed. It is advisable to strictly separate the organizational audit from the supplier audit.

2. Liaison / Reporting Requirements
   A business liaison, such as the supplier's key account manager, should be nominated for the specific audit steps agreed in the initial interview. Reporting should be comprehensive and in line with the standardized audit program. The auditor should refrain from investing any time and effort (unless the supplier has budgeted for this) in assisting with fact-finding within the supplier organization. The objective of discussing reporting requirements with the supplier is to give clear instructions and elicit a set of data and documents that mirror the document structure agreed for the principal firm.

3. General Audit Structure
   Present the structure of questionnaires and further interviews to the person nominated by the supplier. Distribution, processing and return deadlines should be agreed upon. Measures of completeness, accuracy and supplier participation may be defined to ensure that the data returned are sufficient for analysis. These performance measurements should be approved by both the supplier and local management (or the BCM coordinator).

4. Miscellaneous
   Clearly link the supplier audit to the performance levels stipulated by the service level agreement. It should be jointly agreed between auditor and supplier that the auditor is acting on behalf of the principal, and that the supplier is expected to conform to the principal's audit effort.

Interview guidelines may be designed to cover a high-level agenda (as in the examples), or be more detailed. The latter approach may be beneficial when the auditor is supervising an audit team working in a decentralized manner. Alternatively, a workshop or training session for interviewing may be appropriate to educate all audit personnel as to the objectives and techniques to be used. However, too much detail may be counterproductive, as this may impede flexibility and discussions between auditor and auditee. "Instruction overkill," telling the auditor the precise verbiage for questions, the timing of the interview, and even actions such as "Step 1: breaking the ice"[32] tend to create insecurity or an unnecessarily rigid atmosphere that does not encourage the free exchange of thoughts.

[32] the author's personal experience does indeed include instances where interview guidelines were worded along the lines of "do:, say:, act like:". Despite the natural wish to insert some uniformity into the activities of auditors, this level of granularity is unlikely to provide any benefit.

Example of BCM Questionnaire

The sample questionnaire printed below illustrates how a predetermined set of questions may be derived from the standardized audit program in section II. In this example, some aspects of detailed business continuity planning (area 6) are addressed by a combination of open textual questions and multiple-choice questions. The choice of question is limited for reasons of space, and it is likely that more exhaustive questionnaires would be applied in a real life audit situation.

Although these questions may, at first sight, appear somewhat superficial, it should be noted that the auditor's primary interest is to confirm parameters such as existence, depth, documentation and audit trail rather than operational issues pertaining to facilities management. In other words, the auditor should refrain from asking "why" a certain function within facilities management is performed in a specific way. This is one reason for the rather unspecific nature of some questions which avoid addressing specific business processes, as would a business impact assessment.

Detailed BCM Questionnaire
Subset: Facilities Management

The following questionnaire is part of the audit plan for Electronic Widgets Ges. mbH, Austria. The questions address issues of business continuity management in the area of facilities management. The main objective is to determine whether adequate business continuity planning has been undertaken for all aspects of facilities management, and whether the organizational environment is conducive to maintaining and managing such plans in an appropriate manner.

Please answer all questions to the best of your knowledge and abilities. If any question should be unclear or ambiguous, please contact the audit team on extension 1234.

1. Who is responsible for continuity of functions and tasks assigned to the facilities management department?

   a) Head of Facilities Management
   b) BCM Coordinator
   c) Senior management (e.g. CTO, COO)
   d) External facilities management company

2. Are the functions and tasks to be performed by the facilities management department documented in a comprehensive manner?

   a) all functions documented, with a clear description
   b) some basic functions documented, secondary functions not documented
   c) no documentation, functions performed through practical experience only
   d) no documentation, function outsourced (partially or completely)

[These two questions are samples for typical multiple choice questions. Note that the answer choices are always limited to four, and no "open end," i.e. "other, please specify" has been included. In any questionnaire, these questions would be limited to distinct factual items that are usually technical in nature. For instance, there is a limited number of methods for alternative power supply solutions. When trying to establish which method has been used, the auditor might apply a standard multiple choice question like the ones above].

3. Does the facilities management department have a disaster / crisis plan ? If yes, please state the areas and general functions covered by the plan. (use additional sheets of paper if needed).

[When dichotomous questions are included in the questionnaire, one of the alternatives (either yes or no) will usually require an explanation. In this example, the decision "Y/N" has been combined with an

instruction to provide more detail. Some questionnaires would separate the question by branching off "if yes, go to Q 27". When using the standardized audit program, this type of question will constitute a significant proportion of the questionnaire, as the auditor is bound to test the existence (step 1) and then the quality of certain items in area 6].

4. Please state briefly how the HPVAC functions for the corporate premises would be recovered in the event of a disaster. (use additional sheets of paper if needed).

[An open textual response is encouraged, also testing contents against preceding "closed" questions. The auditor may find that in the answer, some indicators are present that validate or invalidate earlier answers to dichotomous or multiple choice questions. It is advisable to formulate open questions in this manner to give the auditee as much opportunity as possible to actually express complex facts and their interdependencies. However, analysis of these answers may be cumbersome].

5. Please complete the following table by giving your professional opinion on all of the items. Depending on your experience and knowledge of the organization, indicate the level of business continuity related plans, actions and management procedures as indicated in the table. If you are unsure or unaware of some of the items listed, please mark "not sure". If the item appears to be irrelevant to your operation, please mark "not applicable" and provide a brief explanation.

Audit Item	1	2	3	4	5	?	not applicable because...
Backup power supply							
SLA with power company							
Surge protection							
Power line conditioning							
UPS units							

1= nonexistent, 2=planned, 3=partially implemented, 4=implemented, 5=optimal, ?=not sure

[Tabular questions compressing several multiple choice items are used here to check on several related items. They are applied here to save space and reduce boredom for the auditee. While these questions would appear ideally suited to "clusters" of technical data, the answers are often deficient in terms of quality, as respondents either tick "5" for all items to be "best in class" (thus defeating the object as becomes clear from the above example) or choose to err on the cautious side by ticking "not sure" far too often. The auditor should factor in an increased effort for clarifying and spot checking the answers. The – typically technical – subjects addressed by the tabular format do require such detailed checks].

The example, albeit very brief, shows that specific areas or business processes may require some in-depth knowledge on part of the auditor or audit team. As has been emphasized in the comments to the sample questionnaire itself, the main purpose of the audit would be to ascertain the existence of an adequate procedure for supplying power in the event of a disaster, rather than being trapped by asking what the actual procedure entails. If such detailed tests for adequacy are required, technical experts should be consulted prior to distributing the questionnaire (to ensure that the questions elicit the desired information) and during analysis of the returned questionnaires (to give an opinion about the adequacy).

Pitfalls and Known Difficulties

The sample interview guidelines indicate that for the auditor, it is less important to work through a standardized set of questions. The primary objective of interviewing management and the BCM coordinator is to obtain a realistic understanding of the overall situation. In addition to the example shown above, tentative topics may be agreed for further meetings, as shown in the "Miscellaneous" sections of the interview guidelines.

The design of questionnaires is rather straightforward, as sets of related questions may be derived directly from the standardized audit program. The depth of these questions in each area is determined by their decimal number (see section II). It is thus possible to use any number of audit items from the standardized program to either obtain a high-level understanding or "drill down" if a particular point merits closer inspection.

Expectation Gaps

Despite the formal approach taken in the example above, it is the auditor's experience that determines success or failure. Conducting interviews is as much a function of "chemistry" as it is dependent upon preparation and guidelines. More often than not, the auditor will find that after the first interview on site, the situation appears incompatible with what was discussed in earlier phases of the audit planning stage, at corporate level. "Miscellaneous" factors influencing the hands-on work on site may be discouraging, for instance the presence of government regulations or a lack of BCM awareness in the general management culture of the country where the audit takes place. In the extreme, auditor and auditee may discover that they are "talking in different languages." In most cases, there will be several individuals – even in such an adverse environment – who have recognized the need for business continuity management and its corporate standardization. Hence, one of the main purposes of initial interviews (as opposed to merely distributing questionnaires) is to test awareness and identify key individuals who actively support both the audit and resulting BCM initiatives. In joining forces, the auditor wields considerable power by dint of his / her external mandate to impartially evaluate the situation at hand.

Obviously, the interviews are the main determinant of subsequent audit activities. If the ground has been prepared for these further steps, the auditor is likely to experience a lot more support and understanding. The secondary function of the face-to-face interview stage is therefore to bridge the expectation gap that may open between auditor and auditee, as expressed by the following typical statements[33]:

[33] All of these statements are real, but anonymized.

Auditor	Auditee	Explanation
"I'd like to go through the processes that you've got for BC coordination and compliance"	"Well, our backup and tape storage has just been modified to make sure we don't lose data anymore."	The auditor is thinking in terms of a mature BC department or staff unit reporting to senior management, perhaps the CIO or CFO. The auditee thinks like most auditee organizations in that country.
"Can I see a classification table for your critical business functions?"	"The most critical task is manufacturing. We have taken precautions to make sure the production lines are operating at all times."	Again, the auditor is thinking in terms of an abstract classification, whereas the auditee believes that "critical function" is equivalent to "core business." Which is not wrong, but it addresses only part of the auditor's question.
"Is there a chart that shows roles and responsibilities in the crisis command center?"	"We're just five people here, and we get on extremely well. There is a lot of informal communication, and everyone generally knows what the others are doing."	The auditee thinks about formal job descriptions and the informal way jobs are being handled under normal conditions. The notion of a crisis command center is obviously unknown.
"Are the plans for crisis PR published on your corporate Intranet?"	"Our PR department does that. They report to the CEO. We've tried to talk to them, but they think they know best how to handle it."	The auditor is thinking of an integrated PR strategy for crisis situations. However, it seems that the corporate PR department is jealously guarding their secrets, encouraged by senior management.
"What's your liaison with the local authorities?"	"Some time ago, the fire brigade took a walk through and made recommendations. When we built the new annex, the police came round and suggested an alarm system. There was a local surveyor once who gave us a list of modifications that we had to do before the new annex got approved."	The auditor is thinking of area 10, civil defense and a typical partnership between the organization and public authorities. The auditee has experienced isolated instances of external involvement (bothersome) and wishes to dodge the issue. This clearly represents a fundamental misunderstanding re. area 10.

Cultural Divergence

The expectation gaps shown in the table above are a major problem in auditing business continuity. As a relatively recent discipline, BCM appreciation significantly diverges between cultures and regions throughout the world. Furthermore, as indicated in the example in Chapter 1, a tendency has emerged to view the Anglo-American approach towards business continuity as the predominant paradigm. Leaving aside the language dimension, strict adherence to what the auditor is used to in terms of BCM may not always be the best strategy. In order to bridge the resulting expectation gaps, it may be more beneficial to first obtain an understanding of the general set of values or management thoughts at the center of which are different notions of BCM. Consider the following examples:

- in the German-speaking countries, namely Austria, Germany, and Switzerland, business continuity stems from a time-honored technical and engineering tradition. Power plant engineers, for instance, often exhibit an outstanding level of BC knowledge and practical experience[34], but they would never refer to what they are doing as "business continuity"[35]. Conversely, business consultants and experts from other disciplines have adopted the term "BCP" or "BCM", but the terminology has mutated. Using the very same words almost inevitably means talking about two very different things, as the second group mentioned means IT disaster recovery when talking about BCM.

- in some Asian countries, the fundamental concept of business continuity – to maintain a reasonable level of performance and operations in a critical situation – is an alien one[36]. Local culture, influenced by centuries of sophisticated social and commercial development, prefers a cyclical perspective on life and business: after the disaster or crisis, something new will emerge, and it will be dealt with accordingly. The intrinsic value of restoring "what was" to exactly the same state is questioned or even rejected.

- certain countries and regions have experienced, or are experiencing, periods of major unrest and a constant state of force majeure conditions. Civil war, natural disasters, or environmental conditions have led to an overall business model that is based upon improvisation. Business continuity, in these settings, is not perceived as a value per se, as nobody is able to continue their normal operations anyway. Economies undergo constant and unexpected changes hitting all businesses alike. In a climate of political, economic or natural instability, business continuity – to continue at all – has been internalized to the point where planning for it might be seen as ironic.

The auditor will need to reconcile these cultural backgrounds with the overall audit objectives as defined in the scope statement. As will be described below, this may have an impact on analysis and reporting. It may not be advisable to force the issue when performing on-site audit activities, particularly where foreign cultures are clearly divergent from the auditor's "home" culture, or from the original audit idea discussed with a clear perspective on the corporate culture prevailing in the "home" country. As continuity has such a diverse understanding and many nuances in different world regions, some adversity may build up if a single method or audit plan is applied regardless of the cultural background experienced on site.

Specialist Expertise

When conducting the audit, the question of extending the auditor's skills to adequately cover all areas will undoubtedly arise (*see Chapter 1*). Various strategies have been suggested to overcome this latent difficulty:

- assign corporate personnel to support the audit, including travel to the audit location to support special tasks
- assign local staff where available to provide specialist expertise
- assign local third party experts to special tasks

---

[34] cf. as representative examples the writings of Kautz between the 1980's and 1997.

[35] this is evident in many recent endeavors, for instance the Romanian national network project, cf. Delamarian and von Rössing (1997).

[36] cf. Craig (2000) from a local perspective.

While these steps usually improve the overall skill-set at the auditor's disposal, the reliability and validity of expert judgments and opinions may vary, and the assessment of results thus obtained is a double-edged sword. Experience shows that there may be a trade-off between efficiency of specialist resources and their adherence to the objectives of the audit. The auditor cannot always judge the quality of work provided by specialists (otherwise there would be no need for them in the first place). Corporate personnel specializing in BCM may prove ineffective in remote countries where they have had no experience. Local staff may not have the necessary skills that the auditor requires. Third parties may follow their own agenda and maximize their involvement rather than pursue the audit objective. There are many caveats that the auditor will have to consider prior to enlisting experts. In the great majority of cases, the seemingly trivial question of language emerges as the single most important criteria for selecting experts.

Problem	Recommended Solutions
Specialist (external) entertains different ideas on BCM audit	Prior agreement on audit scope and planning, sign-off by external specialist
Specialist (external) extends involvement to maximize profit	Limit timeframe and budget prior to commencing on-site audit, include limit clauses in service contract. Agree on milestones and deliverables on a pay-per-milestone basis.
Specialist (internal) has difficulty in adjusting to local culture and conditions	Appoint local "counterpart" and regroup teams where appropriate.
Local staff do not have required skills or experience	Bring in corporate specialist staff as necessary. Support local training and awareness measures.
Specialists (internal or external) have difficulty in adjusting to audit requirements	Provide knowledge transfer on audit, not BCM. Reiterate audit framework and objectives.
Local culture does not support audit (for various reasons)	Shift focus from audit to consulting aspects of the project. Research background and reasons for adversity or lack of support. Readjust audit objectives if mitigating circumstances exist, after consulting with audit sponsor.

The above table shows some typical problems encountered when enlisting expert help on an international basis. While the "recommended solutions" can only be indicative of the choices the auditor may have, they nevertheless point to the fact that a rigid approach towards pushing a uniform agenda may sometimes not be enough to accommodate unforeseen circumstances and difficulties. It is suggested that the auditor should remain reasonably flexible in terms of reacting to cultural, political and third party issues. Where in doubt, it may be advisable to discuss the points in question with the audit sponsor, the audit committee or local management, depending on the nature of the problem.

While business continuity skills are comparatively easy to ascertain, the overall capability in terms of accompanying and supporting a BCM audit is much more difficult to assess. It is suggested here that, in addition to traditional criteria applied to the selection of staff and consultants, the auditor should secure the assistance of a managerial "counterpart" very early in the process of scoping and planning the overall audit. In addition to satisfying the criterion of "four eyes" reviewing the scope and individual plans, this practice combines external knowledge with the auditor's internal network within the auditee company[37].

Summary

- Prepare a schedule of on-site activities, namely interviews, questionnaires and administration. Within the agreed timeframe and budget, minimize disruption to the auditee organization, without compromising the agreed audit objectives.
- Take into account the administrative overhead, logistics overhead (e.g. preparation of travel, familiarization) and additional time buffers for clarification of information gathered. A significant percentage of audit time will have to be allotted to "housekeeping" tasks such as status meetings or interim reporting to stakeholders.
- Determine the proportion of personal activities, i.e. face-to-face interviews and meetings, versus standardized activities such as questionnaires. The allocation of effort to these categories should be driven by the objectives set down in the audit plan.
- When using questionnaires or other mass information-gathering devices, bear in mind that they have a significant degree of finality – once distributed, there is limited scope for re-runs or rework.
- Questionnaires and similar instruments, e.g. software-based data gathering and survey mailings, should be directly derived from the standardized audit program to ensure compatibility with the audit plan.
- Avoid known pitfalls and difficulties such as expectation gaps arising from the local environment, cultural divergence, or the lack of specialist expertise often found on site.

[37] assuming that the auditor is an employee of the universe of companies and subsidiaries being audited. This idea works vice versa, of course, if the external auditor appoints an internal "old hand" to benefit from their experience.

3

ANALYSIS

Chapter 3 outlines methods of analysis that you can use to arrive at a well-founded audit opinion. As a financial or internal auditor, this chapter will allow you to evaluate your findings and to avoid time-consuming detail when reviewing the BCM process. As a business continuity manager, you will find Chapter 3 a useful tool for looking at any given part of a BCM process and for comparing findings against your own experience and best practices.

Structuring, analyzing, and interpreting data from a business continuity audit is, arguably, the most challenging phase of the audit. In contrast to other activities, the analytical process directly reflects the auditor's BCM knowledge and requires professional judgment as well as consideration of lateral interdependencies. The data gathering stage, as described above, often tends to produce an overwhelming amount of data, particularly for large and complex audit plans. Collectively, the data represents the basis of subsequent reporting, but appropriate analytical methods must be applied to make sense of it all. In comparison with financial audit and the well-established tools for data analysis used in conjunction with its sub-disciplines, the BCM framework is somewhat less developed. The direct application of known tools and methods from financial analysis would not yield any meaningful results in a business continuity context.

There are a number of fundamental differences between BCM audit and other forms of auditing and reviewing. First, data gathered tend to be qualitative. Quantification is difficult at best, and often arrived at in an indirect manner. Second, the quality of business continuity as reviewed in a corporate entity is speculative by definition. Only in an actual crisis or disaster situation could the full validity of a business continuity plan be correctly and accurately assessed. Accordingly, analytically strong statements indicating "correctness," "completeness" or "adequacy" are extremely difficult to ascertain in conjunction with any BCM audit. Likewise, the opposite notion of "incorrect," "incomplete" or "inadequate" are subject to challenge unless the BCM approach taken by the entity under review has been tested (and failed) in a critical situation. Thirdly, the normative strength of known standards and regulations is far less developed than financial audit standards such as IAS, SAS or GAAP. Hence, measuring factual findings against an ideal-typical world – as found in many accounting standards – is usually impossible.

Leaving aside such issues arising from general methodology, the very simple obstacle to analyzing BCM in general is that the auditor does not normally possess the knowledge and experience to perform detailed analytical steps at a consistent depth. In most cases, there will be areas where the auditor can contribute extensive prior experience, as would be expected from a CPA for finance, a CISA for IT-related matters and so forth. For other aspects of analysis, it is inevitable that the auditor will have restricted knowledge, or indirect knowledge through the use of specialists. For instance, an experienced financial auditor may feel quite comfortable with calculations of business impact and their adequacy in terms of materiality, but a lot less comfortable with judging the plausibility of a backup data center environment. Conversely, an experienced IT auditor will view the audit of a complex multi-site e-business environment as straightforward, but may have difficulties when faced with emergency authorization rules for cash disbursements.

The following sections present a step-by-step approach towards structuring and analyzing typical data gathered in the course of a BCM audit. While some of these steps are generic, others specifically refer to the field of business continuity and its "fuzzy" context. The standardized audit program (see section II) is then used as an example to illustrate the practical application of analytical methods and tools.

Summarizing Interview Results

Having conducted several face-to-face interviews, supported by questionnaires and other data gathering instruments, the information thus obtained requires sorting, grouping and condensing. Depending on the total size of the audit program, this process of streamlining data will be more or less formalized. Given a limited audit scope, there is no immediate need for applying a formal process to elicit comparable information from a large number of interview notes. At the other end of the scale, in a multinational audit with several auditors, it is evident that some form of calibration is needed to ensure clear and unbiased interpretation of what data and information the auditee entities have provided. Summary information compiled from interviews and other data is therefore a prerequisite to answering the overall question behind the audit (*see Chapter 1*).

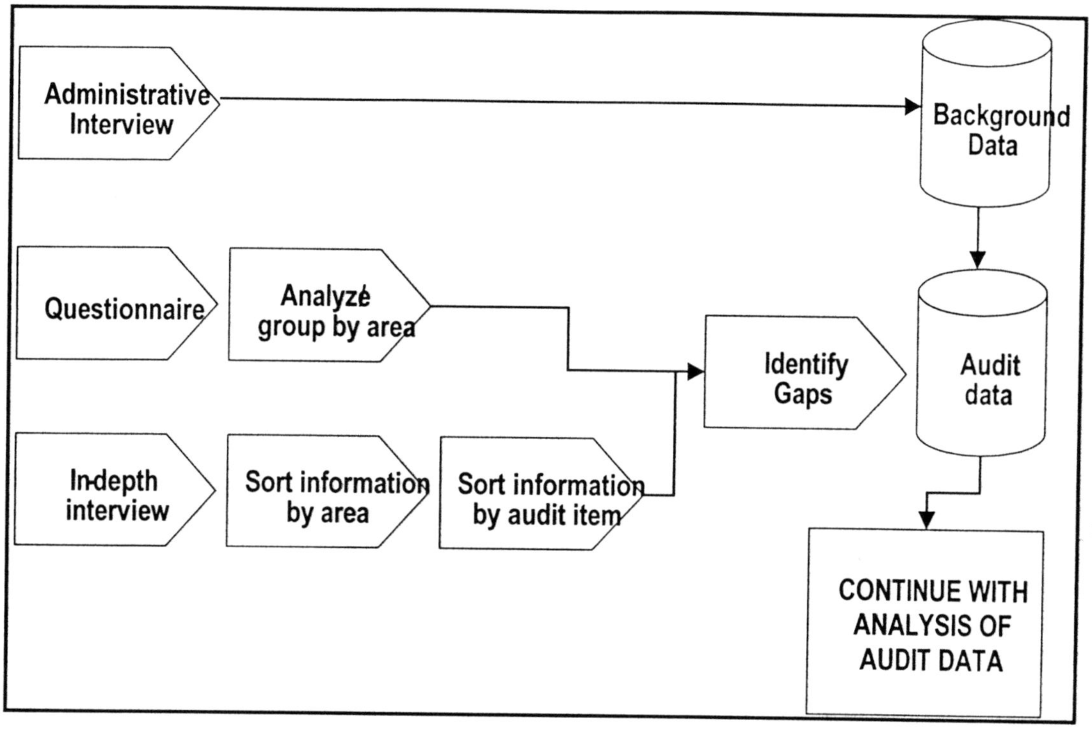

The auditor should distinguish between interviews of an administrative nature and those referring to specific audit areas. While the former may give an indication of the organizational background in the auditee entity, it is the latter type of interview that usually provides salient information for the purpose of analysis. As a rule of thumb, any interview that emulates – or is equivalent to – a questionnaire, should be set aside for analytical treatment. Likewise, any interviews relating to individual points or whole sections of the standardized audit program should be formally analyzed to assess the audit item(s) addressed.

Assuming a typical interview relating to a number of audit items, such as technology and infrastructure, it is very likely that the result will be less than structured. Despite the best efforts, the auditee will have strayed from the carefully prepared audit guideline, and incidental information may have been offered which, at first sight, does not have any direct or overt relationship with the "planned topic." The psychological setting of an audit may have prompted any response from aggressive venting to evasive behavior, and the auditor, having duly noted all utterances of the interviewee, is expected to systematically group the information obtained[38]. In larger audits involving teams of – relatively inexperienced – auditors, the phenomenon of seemingly irrelevant or "beside the point" data is often seen as an annoying detail that precludes proper analysis. Even experienced audit professionals who have not been trained in interviewing techniques may encounter similar difficulties. However, these apparent deficiencies may be used to the advantage of the auditor. Any information given in the course of an interview should be regarded as an additional source of wisdom, restricted only by considerations of project control (time and budget, see below). Hence, the seemingly unstructured interview may prove invaluable in deducting information rather than having to ask for it. It is suggested here that the "between–the-lines" information thus gleaned may even be of higher value than the standardized responses to standardized questions.

[38] This phenomenon may be even more difficult to control when the auditor is faced with a different language environment or a different culture. Despite the benefits of "active listening," the resulting loss of control is – at first – a source of frustration.

Step 1: Visualize the interviewee's narrative.

Whatever has been said during the interview is usually straightforward to group thematically. As dictated by human nature, the interviewee is likely to dwell upon a subject unless stopped, depending on how important that subject is to the person being interviewed[39]. The auditor should therefore attempt a rough structuring of the interview narrative in terms of time, for later comparison with other interviews:

- how much time has been spent discussing the subject / topic?
- were there any emotional signs or expressions signalling high importance?
- were there any signs hinting at the interviewee's assessment of this topic (particularly frustration, cynicism etc.)?
- in comparison, how much time did the interviewee devote to the item being discussed, as opposed to the attention span found in higher / lower echelons of management?

In BCM audits, there appears to be a marked tendency for interviewees to transcend the set boundaries of interviews and to verbalize certain issues that, in their own mind, have a high priority. In terms of analysis, this may be a sign of personal fear (what people are most afraid of), personal empathy, or even idealism (the world how it should be, were it not for management). The wary auditor should nevertheless acknowledge that "jumping" from one audit item to another often indicates a useful degree of insight on part of the auditee who is much more deeply involved in the organization and its "hidden workings". Hence, it is up to the auditor to translate organizational reality (!) prior to looking at the audit plan, and not vice versa. If the outcome of an audit analysis stage indicates "did not reply" or "no answer," such marks show that the auditor may have attempted at matching unstructured information with a pre-set audit program, in this case unsuccessfully. Obviously, the auditor has failed to interpret interview results and hence, omitted to "tick the box" - unless the interviewees, in their own words, indicate that an answer is denied or unknown.

As a result, the second step in summarizing interview results is to group the interviewee's narrative into the ten areas (or items thereof). Any direct statement made by the interviewee, while perhaps not pertaining directly to what the auditor had in mind, usually fits into one or another area within the ten areas. The auditor should note how the interviewee arrived at the mental connection between a question and a seemingly unrelated topic, as this may indicate both organizational habit and prevalent thinking within the auditee entity.

Step 2: Group the interviewee's narrative by area (10 audit areas) and item within the area.

The combined results of all interviews will thus form an emerging pattern across all audit areas.

[39] Analytically speaking, the emphasis given by the interviewee, and their attention span, clearly indicates the relative importance of the topic.

Specific items of interest may present themselves when comparing the relative frequency of occurrence, and the auditor's conclusions form an important part of the overall analysis stage.

If and where the resulting "jigsaw" of interview statements shows that some audit items or areas may not have been covered at a sufficient level of depth, additional interviews or clarifications may be necessary. However, given the disruptive potential intrinsic to arranging meetings with auditee representatives, these requests for clarification (as part of the analytical stage) should be used sparingly and with clear reference to the "blank" areas of the audit plan. In practical terms, BCM coordinators or senior managers will be more likely to consent to setting aside further interview hours when given a clear reason why a second round of meetings is essential.

Step 3: Identify gaps in the audit plan that have not been addressed in interviews. Request clarification as appropriate to complete the data gathering phase.

In summarizing interviews, the predetermined level of granularity (as determined by the audit plan, see above) should be taken into account. In order to prepare a map or matrix of interview results, it is essential that all summaries be formulated in a manner that allows tabular representation and comparison with either other corporate entities, or individual audit items selected from the standardized audit program. In a situation where several persons are involved as auditors, the comparability of information elicited from face-to-face interviews may be the decisive factor for analytical purposes. When using questionnaires, which provide a requisite level of standardization, this alignment tends to be less important due to the formalized way of asking questions and processing answers.

Example of Interview Series Summaries

Following the hypothetical audit of Electronic Widgets, the following example presents some rough interview notes whose information content is then analyzed by means of the above-mentioned techniques. It is assumed that the ten standard areas and the standardized audit program have been applied in planning the audit. This means that "sorting" verbal statements into different audit areas is much simpler than having to draw up categories to fit the interviews. However, as apparent from the rough notes, there will inevitably be some remarks that do not easily fit into any category. The auditor should nevertheless keep these in mind for later use.

Electronic Widgets Ges. mbH
Vienna, Austria

Interview Notes, Facilities Management
Interviewee: Josef MUELLER, Facilities Manager

Facilities Management (FM) seen as an important part of the general business, but underfunded. Common resource problems prevent the implementation of enterprise-wide programs and improvement initiatives Mueller reports to CTO, but that individual has been appointed recently, to succeed a person who had been in the company for 27 years. The new CTO has initiated some cost-cutting projects, notably the reduction of

FM staff
[talked about this for about 15 minutes, sort of an introductory remark by Mueller]

> *[Mueller, apparently a veteran within the organization, seems to be concerned about the strategy pursued by the new CTO. From the interview notes, it is difficult to say whether this is just a general complaint that any facilities manager might voice, or if this represents a BCM issue.]*

The general economic development has led to a lot of "problems with buying new technology." Security has been neglected, and some fire protection regulations have not been followed. Recently, an independent consultant has assessed these issues and found several serious flaws such as sprinklers in decentralized data center areas and computer rooms. The CTO seems unresponsive to these concerns.
Mueller has advocated the installation of a general "very early warning system" [ionization sniffing device ?] for fire prevention, at least for the data center areas. However, no budget has been authorized for this.
There is a regular gathering of FM professionals in Vienna where people talk shop. At the last meeting, they discussed the potential of a major fire in a data center. Mueller appears concerned about the collateral damage which might occur in a typical high rise building with the current fire fighting capability. This seems to be a general concern in Vienna.
[about 10 minutes for the fire topic]

> *[There appears to be a technology backlog. Given that the general opinion amongst Viennese facilities managers is undivided, this item might merit closer attention. Some details indicate that budgetary constraints have led to an irregular situation in which the FM department is prevented from complying with regulatory requirements. Fire protection is always serious, and might indicate a connection with the budget issue. Note that Mueller talked about these detailed points for 10 minutes. This suggests that he is quite concerned about the topic of fire protection.]*

FM have formulated an emergency response plan that describes evacuation procedures, mustering stations and the procedure for head counting. Mueller feels that – given that a couple of tests have been performed – evacuation works reasonably well. However, the diesel generator issue remains open. Despite the evacuation of people, the controlled shutdown of all systems is not guaranteed. Mueller has had an issue with the IT manager who has not responded to his requests for shutdown periods.
Signage, lighting and mechanical locks for the interior of the building are all implemented precisely to specification. Mueller suggested a walk-through to verify. Egress from the building in the event of an emergency is not a problem.
Once before, during a minor incident, Mueller was intercepted by the press upon leaving the building. He feels uncertain about his response, as his statements were printed in an entirely different wording. As a reaction, he has served a general notice forbidding anyone within the FM department to talk to the press if an incident occurs.
Media seem to be unaware of the typical problems associated with a large facility such as the high rise office tower occupied by Electronic Widgets. Mueller pointed out that many of the problems faced during the incident were caused by other tenants, namely the "Wellness Institute" on the top floor having to evacuate their guests. Their statements implicated Electronic Widgets as the main tenant of the building, but the press never checked whether these accusations were actually true.
[about 10 minutes discussion of incident]

> *[Characteristically, Mueller "jumped" between different audit areas. While the procedural part of BCM seems to be well established (and exercised), the interface with the media leaves quite a lot of things open. Mueller appears to have been caught unawares by the press in an unfortunate situation. Likewise, cooperation with the IT department is less than perfect – the reasons to be determined by the auditor.]*

Mueller fervently supports the audit and likes the idea of putting in place a "proper" BCM. He feels that the current initiatives are sketchy at best, and that from an FM point of view, there should be a lot more in terms of maintaining the infrastructure.
He has had talks with the building owner who has indicated a willingness to share in the financing of a "complete" business continuity plan. However, the CTO has rejected the idea on grounds of "more important projects", and Mueller has had to decline the offer by the building owner.
Mueller requests that his concerns be verbalized in the audit report, and emphasized accordingly. His safety and security concerns are – in his opinion – directly linked to building regulations in Austria. For the FM department, Mueller declines responsibility for any incidents as long as there is no budget to properly support him.

[about 15 minutes for audit discussion]

[Mueller has taken the initiative and enlisted the support of the building owner. Given his own – quite sensible – notions of BCM, he has even managed to convince the owner that BCM is an important topic. The CTO may have been insufficiently prepared to recognize the potential for sharing a significant cost burden in this particular case, which should be taken note of by the auditor. The reference to regulations should, as always, prompt further investigation by the auditor, as the rejection of responsibility on part of Mueller indicates that something is seriously wrong.]

(Refer to documentation) Plans produced; deposited with working papers, not yet reviewed. At first sight, the FM part of BCM seems well developed, with some clearly delineated gaps. No forward planning for the next three years.

[This section, to complete the interview notes, just indicates the documents reviewed and collected by the auditor. It is not strictly part of the interview, but is nevertheless useful to show what documentary steps have been taken and what the relevant documents are with regard to interview contents. The auditor should naturally check the above-mentioned documents for any discrepancies or inconsistencies with interview contents, as well as noting supporting evidence.]

[...]

Electronic Widgets Ges. mbH
Vienna, Austria

Interview Notes, Communications Department
Interviewee: Leopold KRANTZ (Head of Dept)

According to Krantz, the comms dept has an integrated communications strategy that is proactively aligned with senior management objectives. Krantz regularly talks with the CEO and CFO to make sure that their intentions are communicated internally and to the public.
Comms is clearly a managerial department. Krantz does not know about technological issues and tends to leave these to the IT manager. FM have not been involved in drafting communications plans and media statements.
[about 10 minutes, technical and FM items about 2 minutes]
Krantz has invented a diary outlining events, press appointments and other items of interest to the comms department. He maintains a database with important media representatives and prides himself on keeping regular contact with them, usually through power lunches or meetings in media circles.
There are no specific provisions for a crisis situation, as this has never happened before. The evacuation incident was not seen as significant by senior management. In the event of a disaster or critical situation, Krantz feels that his experience and the general coherence of the team will allow the comms department to be in command of the situation. Two individuals have more than 15 years of PR experience, and the younger members of the team are being trained on a regular basis.
[approx. 25 minutes]
The notion of a command center has never been suggested by the CEO. The nearby hotel may be able to offer a couple of conference rooms, as Krantz is on good terms with them. Given that Electronic Widgets is a good customer, there wouldn't be any problem in securing space if there were a disaster.
Krantz favors an "all–hands-on-deck" approach if something happens. In his opinion, it is best to have all people show up and then brief them on the situation. He has prepared a presentation to employees outlining how investor relations will be unaffected by the event itself. This has been approved by the CEO and the CFO.
[about 15 minutes including show of presentation]

[...]

The interview notes for both interviewees appear sketchy and seemingly unstructured. As a first step, it is therefore essential to derive the salient facts from what was said, and to highlight their relevance to the ten audit areas. In the above sample notes, the facilities manager and the head of the communications department have been chosen to allow cross-referencing and thus identify other issues by means of deductive logic.

Textual Analysis of Interview with MUELLER (Facilities Management)

The focus for the interview was on gathering information about the following:

- technical infrastructure, HPVAC, fire protection etc.
- known risks and weaknesses in business continuity
- opinion about the ability to survive adverse events
- crisis management capability (theoretical)
- clear strategy for BCM and FM implementation of this strategy
- existence and extent of detailed BC plans
- training, general corporate awareness and FM "buy-in" to a corporate BCM strategy
- Tests and maintenance of plans and related technical measures
- Links to crisis communications and public authorities (the latter in particular)

Facilities Management (FM) seen as an important part of the general business, but underfunded. Common resource problems prevent the implementation of enterprise-wide programs and improvement initiatives

> *[**Area 1**: Project initiation and management seems to be deficient with regard to the FM department. If a proper BCM project were in operation, Mueller would have mentioned this.]*

Mueller reports to CTO, but that individual has been appointed recently, to succeed a person who had been in the company for 27 years. The new CTO has initiated some cost-cutting projects, notably the reduction of FM staff

> *[**Area 1**: Can a BCM project / process be accomplished with reduced levels of staff? This indicates a project management risk, i.e. requiring the facilities manager to do more with less. This will only work to a limited extent before quality of the FM process will begin to deteriorate.]*

[talked about this for about 15 minutes, sort of an introductory remark by Mueller]
The general economic development has led to a lot of "problems with buying new technology". Security has been neglected, and some fire protection regulations have not been followed. Recently, an independent consultant has assessed these issues, and found several serious flaws such as sprinklers in decentralized data center areas and computer rooms. The CTO seems unresponsive to these concerns.

> *[**Area 2**: some very serious risks are evident from this paragraph. The independent review should be scrutinized and incorporated into the audit report. If the CTO is unresponsive, this may be due to several reasons. However, items such as fire protection may indicate non-compliance with regulations or the law, thus requiring more detailed analysis].*

Mueller has advocated the installation of a general "very early warning system" [ionization sniffing device?] for fire prevention, at least for the data center areas. However, no budget has been authorized for this.

> *[**Areas 1, 2, 5, 6**: This detailed remark points to the fact that project funding is insufficient if even essential fire protection mechanisms are not authorized. Moreover, there seems to be at least one flaw in the overall BCM strategy (Area 5) that has prevented senior management from taking the fire protection issue into account. Thirdly, there must be significant risk in the data center areas if Mueller has asked for a technically sophisticated early warning system. This risk may be due to the actual amount of combustible material there, or the value of equipment placed in the data centers may be considerable. The auditor might like to follow up on all these thoughts.]*

There is a regular gathering of FM professionals in Vienna where people talk shop. At the last meeting, they discussed the potential of a major fire in a data center. Mueller appears concerned about the collateral damage which might occur in a typical high rise building with the current fire fighting capability. This seems to be a general concern in Vienna.

*[**Areas 2 and 3**: it appears that the Viennese FM community have neglected to tackle the question of business impact. Furthermore, an isolated risk is being discussed without any visible reference to a systematic risk analysis. Significantly, Mueller failed to name a figure for the potential collateral damage to Electronic Widgets. This suggests that the BIA may not have been conducted in a comprehensive manner. Mueller's colleagues from the professionals circle seem to be in a similar situation, indicating that none of the participating companies are capable of addressing the BIA issue in a coordinated way].*

[about 10 minutes for the fire topic]
FM have formulated an emergency response plan that describes evacuation procedures, mustering stations and the procedure for head counting. Mueller feels that – given that a couple of tests have been performed – evacuation works reasonably well. However, the diesel generator issue remains open. Despite the evacuation of people, the controlled shutdown of all systems is not guaranteed. Mueller has had an issue with the IT manager who has not responded to his requests for shutdown periods.

*[**Areas 1, 3, 4, 5, 6, 8**: If there is an issue with the IT manager, there appears to be a lack of involvement on part of the BCM coordinator. Otherwise, that individual should have addressed the weaknesses in inter-departmental cooperation. With regard to Area 3, the IT manager is obviously unable or unwilling to provide minimum timeframes for controlled equipment shutdown. This shows that the IT manager's BIA or detailed planning is either incomplete or deficient. The auditor should therefore cross-check with the IT department and the BC coordinator to drill down on the questions regarding Areas 1 and 3. Crisis management (Area 4) seems to be well understood by Mueller, who has initiated planning and covered most typical FM issues – at least verbally. Area 8 has been included by means of some successful tests for the initial phases of crisis management. Area 5 merits closer attention, as the autonomous planning effort initiated by Mueller points to a lack of such efforts in other departments. From the wording, it seems that Mueller has decided to take matters in hand, seeing as others are not following.]*

Signage, lighting and mechanical locks for the interior of the building are all implemented precisely to specification. Mueller suggested a walk-through to verify. Egress from the building in the event of an emergency is not a problem.
Once before, during a minor incident, Mueller was intercepted by the press upon leaving the building. He feels uncertain about his response, as his statements were printed in an entirely different wording. As a reaction, he has served a general notice forbidding anyone within the FM department to talk to the press if an incident occurs.

*[**Area 9**: clearly an issue of PR and crisis communications. Mueller's reaction, however, is interesting. It suggests that prior to the above-mentioned incident, there was no direct cooperation between Mueller and the communications department. He was therefore caught unawares, and had to improvise. The auditor should cross-check with the communications manager (interview notes included, analysis below) to find out what happened and why. In any event, the incident suggests weaknesses in Area 9].*

Media seem to be unaware of the typical problems associated with a large facility such as the high rise office tower occupied by Electronic Widgets. Mueller pointed out that many of the problems faced during the incident were caused by other tenants, namely the "Wellness Institute" on the top floor having to evacuate their guests. Their statements implicated Electronic Widgets as the main tenant of the building, but the press never checked whether these accusations were actually true.
[about 10 minutes discussion of incident]

*[**Areas 1, 9**: Another project management issue that highlights a lack of integration with other tenants of the building. As with other neighborhood questions, BCM should have been communicated and at least discussed with the "Wellness Institute". On the PR front, it now appears clear that the strategy pursued by the communications department did not work when used during the press incident. Even worse, Mueller's experience with high-rise buildings was never called upon to find the right answers to press questions.]*

Mueller fervently supports the audit and likes the idea of putting in place a "proper" BCM. He feels that the current initiatives are sketchy at best, and that from an FM point of view, there should be a lot more in terms of maintaining the infrastructure.
He has had talks with the building owner who has indicated a willingness to share in the financing of a "complete" business continuity plan. However, the CTO has rejected the idea on grounds of "more important

projects," and Mueller has had to decline the offer by the building owner.
Mueller requests that his concerns be verbalized in the audit report, and emphasized accordingly. His safety and security concerns are – in his opinion – directly linked to building regulations in Austria. For the FM department, Mueller declines responsibility for any incidents as long as there is no budget to properly support him.
[about 15 minutes for audit discussion]

[All areas: this is a warning that the auditor should use as a basis for further activities. It appears that the FM department and Mueller have a clear understanding of what BCM means and what the current situation is within Electronic Widgets. Mueller, while trying to use the leverage that an external auditor introduces to the BCM question, may be an ally in identifying and formulating further audit findings. Clearly, the auditor should avoid being instrumentalized by any single department. However, in this particular case, the notions and ideas put forward by Mueller seem sensible, well-founded and supported by objective BCM knowledge and experience.]

(Refer to documentation) Plans produced; deposited with working papers, not yet reviewed. At first sight, the FM part of BCM seems well developed, with some clearly delineated gaps. No forward planning for the next three years.

Textual Analysis of Interview with KRANTZ (Communications Department)

Issues to be covered as above, from a communications point of view. The objective of this interview was to determine BCM capabilities and strategic support of BCM-related communications.

According to Krantz, the comms dept has an integrated communications strategy that is proactively aligned with senior management objectives. Krantz regularly talks with the CEO and CFO to make sure that their intentions are communicated internally and to the public.

*[**All Areas:** as would be expected from a professional communications manager, this statement does not directly cover any BCM issues. The auditor is well advised to verify that the "intentions" are verbalized, documented and in line with proper business continuity management. The interview with Mueller suggests otherwise, but any misunderstandings should be clarified prior to noting an audit finding.]*

Comms is clearly a managerial department. Krantz does not know about technological issues and tends to leave these to the IT manager. FM have not been involved in drafting communications plans and media statements.
[about 10 minutes, technical and FM items about 2 minutes]

*[**Area 1:** the distinction between "technical" and "managerial" departments implied by this statement may be correct in an organizational sense. However, in terms of BCM, the communications manager apparently expects the IT manager to actively cover communications for technology issues. The auditor should check whether this sharing of tasks across departments has produced the desired results. Interestingly, Krantz claims that Mueller has been involved in drafting comms plans and media statements. In the light of the press incident mentioned by Mueller, the auditor should drill down and check in detail what was done, how it was documented and how both sides agreed to proceed in the event of adverse media exposure.]*

Krantz has invented a diary outlining events, press appointments and other items of interest to the comms department. He maintains a database with important media representatives and prides himself on keeping regular contact with them, usually through power lunches or meetings in media circles.
There are no specific provisions for a crisis situation, as this has never happened before. The evacuation incident was not seen as significant by senior management. In the event of a disaster or critical situation, Krantz feels that his experience and the general coherence of the team will allow the comms department to be in command of the situation. Two individuals have more than 15 years of PR experience, and the younger members of the team are being trained on a regular basis.

*[**Areas 4, 5, 6, 7, 8, 9:** Krantz has not addressed crisis management. He strictly adheres to senior management's (erroneous) view that the evacuation / press incident was insignificant. There is no BCM*

strategy, as Krantz is confident that he will be able to improvise in the event of a crisis. Detailed plans do not appear to exist. Awareness on part of Krantz (and senior management) is very limited. Tests have never been envisaged. The overall PR aspect is justified in terms of employees' experience, but no statement was made about PR in a crisis. Krantz suggests that "looking good" towards senior management and external press representatives is important, but the interview statements do not address BCM. The auditor should therefore liaise extensively with senior management to verify the detailed points made by Krantz. Inasmuch as Electronic Widgets should be represented advantageously, this overall objective may need to be translated into much more detailed continuity and communications plans which acknowledge departmental concerns].

[approx. 25 minutes]
The notion of a command center has never been suggested by the CEO. The nearby hotel may be able to offer a couple of conference rooms, as Krantz is on good terms with them. Given that Electronic Widgets is a good customer, there wouldn't be any problem in securing space if there were a disaster.

*[**Areas 4, 9**: Once more, Krantz specifically relates to the CEO's opinions and instructions. From an audit point of view, BCM sets some objective criteria for crisis management and related PR activities which are independent of the CEO's immediate instructions. These objective requirements appear to have been ignored. As a result, planning for elementary items within Area 4 has not been done. It is interesting that Krantz mentioned Area 4 rather than Area 9. Again, this may indicate to the auditor that corresponding communications strategies and documentation should be verified in detail].*

Krantz favors an "all-hands-on-deck" approach if something happens. In his opinion, it is best to have all people show up and then brief them on the situation. He has prepared a presentation to employees outlining how investor relations will be unaffected by the event itself. This has been approved by the CEO and the CFO.
[about 15 minutes including show of presentation]

[Area 9: This statement constitutes an audit finding. Regardless of the "all-hands-on-deck" or "stay at home" statement, the presentation does not address the central issue of crisis communications. Krantz pre-empts the effects of a disaster by suggesting that a pre-fabricated message – consciously ignoring the situation – can set things right. The auditor should check if any further steps have been taken to inform the public and other stakeholders. While approval by the CEO and CFO are undoubtedly necessary for all steps taken in Area 9, the auditor should make sure that the CEO and CFO are informed about the overall communications plan, and the objective criteria set by common BCM standards].

The textual analysis of the two sample interviews shows how the seemingly unrelated statements made by both interviewees illustrate a number of shortcomings in the corporate BCM project at Electronic Widgets. However, the "reading between the lines" approach adopted by the auditor should not give way to speculative conclusions that may be unfounded. In the above example, the first impression formed by an external observer might be that the facilities manager, being somewhat frustrated, has taken to running autonomous initiatives for BCM whilst being very critical of senior management. The communications manager, on the other hand, is portrayed as an individual who is less than interested in BCM, and sets clear priorities on maximizing his own advantage and position vis-à-vis senior managers.

While these observations may provide some insights on the interviewee's personal opinions and feelings, they are entirely irrelevant to the auditor, whose immediate sponsor is the company, and therefore senior management. The factual statements – as sharply opposed to the emotional content of the interviews – are quite clear in stating that several initiatives have been taken in the facilities management, but none in the communications department. The central question to be answered by textual analysis is how BCM areas are affected by the interviewees' statements, and which further steps should be taken to verify this. An impartial approach dictates that neither the facilities manager's frustration nor the communications manager's focus on being instructed by senior management must be taken as constituents of audit findings.

The grouping of interview statements into different categories is therefore a first step to extract as much objective and factual information as possible from an otherwise unstructured set of notes. Interpretation of these facts should be done at a later stage in the analysis process. "Jumping to conclusions" should be avoided, as illustrated by the above example. The auditor might find that other departments or managers provide additional information about facilities management and communications that offer new alternatives for interpretation, thus rendering the first impression obsolete. Only when all interviews and subsequent in-depth audit steps have been performed will the auditor be able to impartially validate the situation.

Example of Gap Analysis

Once the interview series has been completed, many audit areas will have been addressed by many interviewees. The next step is to consolidate interview results by area and identify any further interview or clarification requirements. These "gaps" in information gathered will prompt additional audit steps. It is suggested that a tabular representation in three successive steps be used to visualize what is to be done next:

- **Step 1**: draw a matrix of departments against audit areas and indicate which department made any statements in which area(s). Identify "blanks." Verify if these blanks are justified, for instance where a department has no apparent link to the audit area concerned. Decide on further interviews or other audit activities for those cells in the matrix that should not be blank.
- **Step 2**: highlight department/area statements that indicate risks or possible audit findings. Decide on further audit steps or clarification requests for these.
- **Step 3**: prepare a follow-up matrix that groups further action and interviews by department. The result should be a list of planned activities to be presented to the department manager(s) respectively.

Department / Area	1	2	3	4	5	6	7	8	9	10
Facilities Management	☐	☐	☐	☐	☐	☐	☐	☐	☐	☐
Communications	☐	x	x	☐	☐	☐	☐	☐	☐	x
Marketing	☐	x	☐	x	x	x	x	x	x	x
Finance and Accounting	☐	☐	☐	x	x	x	☐	x	x	x
CEO	x	☐	x	☐	x	x	x	x	x	x
CTO	☐	x	☐	x	x	x	x	x	☐	☐
Manufacturing	x	☐	☐	☐	☐	x	x	☐	x	x
...	...									

The table shows how interviewees responded, grouped by audit area. Areas mentioned, indirectly referred to or implied are symbolized by a tick mark, whereas "blanks" are denoted by an "x." The example shows that some blanks are well justified (e. g. manufacturing and PR / communications [9]). Other blanks may require further investigation (e. g. CTO and risk evaluation [2]). The tabular representation of interview results is therefore a useful instrument for a completeness check as well as identifying BCM areas not covered by the departments that would be expected to be involved to a degree.

Department / Area	1	2	3	4	5	6	7	8	9	10
Facilities Management	□	□[1]	□[2]	□	□[3]	□[4]	□	□	□[5]	□
Communications	□[6]	x[7]	x[8]	□[9]	□	□[10]	□	□	□[11]	x
Marketing	□	x[12]	□	x[13]	x[14]	x[15]	x	x[16]	x	x
Finance and Accounting	□	□	□	x[17]	x	x[18]	□	x[19]	x	x
CEO	x[20]	□	x	□	x	x	x	x	x[21]	x
CTO	□	x[22]	□	x[23]	x[24]	x[25]	x[26]	x[27]	□	□
Manufacturing	x	□	□	□	□	x[28]	x[29]	□	x	x
...	...									

1: FM dept, risk evaluation and control – check documentation on fire protection and building safety, verify other risks, physically inspect facilities
2: FM dept, business impact assessment – conduct additional interviews to verify BIA method
3: FM dept, BCM strategy – check alignment with CEO / CTO and overall strategy as documented
4: FM dept, detailed planning – verify if plans are compliant with strategy and management guidelines
5: FM dept, PR and comms – check in detail if any conflict with the comms dept persists and verify documentation
6: Comms dept, project management – verify understanding of BCM project, check CEO / CTO instructions to department, review project documentation
7: Comms dept, risk evaluation and control – review documentation, compare with risk assessment conducted by other departments
8: Comms dept, business impact assessment – obtain BIA documents and validate against other departments' BIA results. Cross-check with finance department
9: Comms dept, crisis management – review documents and suggested steps, conduct further interviews as necessary. Verify the interface between Area 4 and Area 9.
10: Comms dept, detailed planning – specifically check the BC plans for the department and their feasibility. Interview department members to establish experience where written plans do not exist.
11: Comms dept, PR and crisis communications – verify all documentation, presentation materials and senior management approvals. Peruse correspondence and cross-check with CEO.
12: ...
13: ...

The second table shows how high risks and medium risks may be identified for each department and the audit areas covered or omitted. While the first table has not revealed any "blanks" that would require additional exploratory interviews – for instance, if one area had not been responded to by any of the interviewees – the second table highlights several risks arising from the fact that a particular department or individual did not verbally address an area, e.g. CEO and Area 1 (project management). Note 20 might therefore recommend an interview to clarify whether the CEO has delegated BCM project management to the CTO, or simply omitted to provide appropriate senior management sponsorship to BCM as a whole. Likewise, note 21 would recommend further inquiries as to the role that the CEO intends to play with regard to public relations and crisis communications (some CEO response is likely to be demanded by the media and the general public).

The last step in grouping the results is to summarize the actions resulting from the notes listed above, and to convert them into a schedule for each department. This will enable the auditor to agree on dates and times for interviews, delivery of documents, and deadlines for clarification of specific points. If and when the standardized audit program is used (see below and section 2), the process of extracting information from interview notes and grouping it is considerably simpler, as the questions and points to be covered for each audit area are more clearly defined. In this case, the auditor can assess risk and the need for further action as a function of the audit questions answered or answered incompletely, and subsequent follow-up actions are suggested for each question.

Documentation

Depending upon the scope and magnitude of the audit, the level of documentation on both sides (auditor and auditee) is likely to be extensive. It is recommended that a systematic approach be taken that incorporates the audit areas covered, proper cross-referencing and clear guidelines that define the degree of detail. Storing and classifying a vast amount of documents may prove counter-productive if the scope and detailed plan for the audit have a different objective. As outlined in Chapter 1, any documents or working papers included should have a direct bearing on the overall question to be answered by the audit.

Structured documentation techniques are similar to those used in financial audit. There is no single solution to this problem, although smaller audit projects may not require an in-depth documentation structure. For larger audits, the use of an appropriate document management software package is recommended. Leaving aside typical difficulties of access control and confidentiality[40], electronic document management[41] may assist in streamlining the process of assembling and categorizing the work papers as well as documents received.

Any documentation received should clearly refer to the ten areas of the BCI / DRII Professional Practices to facilitate later categorization and cross-referencing. When applying the standardized audit program, this approach ensures convergence between areas covered and documents requested. Given the diverse fields of interest presented by each area, some typical documents and materials may be taken as a starting point for collecting further audit evidence. In order to clearly distinguish between documents providing evidence, those giving an illustration of a fact or set of facts, and those indicating planned activities, the overall documentation system used should indicate the difference with regard to the audit item addressed.

The following paragraphs show some typical documents that the auditor might request when performing the review. Some of these may require further investigation or supporting evidence, whereas others are more suitable for giving high-level information and an overview of complex facts.

Area 1: Project planning and initiation

The auditor should obtain all documents related to project planning and management. These might include project charters, statements of work, initial cost estimates and related correspondence. The overall objective is to identify the approach taken by management and initial supporters of the BCM project idea: if one or more individuals have put forward the idea of embarking upon a business continuity project, the way in which this plan has been communicated to management is as important as the reactions and subsequent decisions reached. It is the auditor's responsibility to determine whether the resulting BCM project has a sound basis or not. Typical documents include:

- Initial project proposal

[40] The reader is referred to the literature on IT security and authorization concepts for further information. For the purpose of this book, it is assumed that the audit team is a homogeneous group with implicit shared trust. or that the documentation is in the hands of a single auditor.

[41] A wide choice of commercial products is available; some BCM packages offer an integrated, albeit limited, document management functionality.

- Management presentation
- Statement of work
- Work breakdown structure
- Project plan (critical path)
- Initial project budget estimate
- Initiation phase communications (memoranda, minutes)
- References to BCM projects in other parts of the organization
- Presentations by consultants
- etc.

It should also be noted that the planning approach and methodology used to finalize the BCM project are subject to review. The auditor should compare these with recognized standards[42] to ascertain whether the project plan in its final form is feasible or deficient in some aspects. For instance, a set of documents indicating an underfunded project with far-reaching objectives might lead to the conclusion that subsequent phases cannot be completed in detail or are, by definition, flawed.

Area 2: Risk Evaluation and Control

Documents on risk evaluation and control should encompass both internal and external risks to the organization under review. There is a wide spectrum of risks that may have been addressed in the immediate environment of the entity or company, and the auditor should focus on the general approach taken towards identifying, evaluating and mitigating these risks. As a rule, risks that people are immediately aware of are often seen as critical, notwithstanding their statistical significance and occurrence. In order to verify the current set of risks, the auditor might consult historical data or independent sources. It is important to note that the overall approach and related procedures are to be audited. Whether the auditor agrees or disagrees with the detailed assessment of an individual risk is usually less important.

As a direct result of the requirements set down in Area 2 of the BCI standard, the level of documentation for risk evaluation and control is usually more detailed than the somewhat generic material on project management. Typical documents requested by the auditor might include:

- Statements on risk management and the overall notion of risk as seen by the organization
- Results and reports of previous risk analyses, including the methodology applied
- Materials documenting risk awareness, i.e., the probabilities and consequences connected with particular risks
- External sources used and their credibility, as well as the interpretation of these sources
- Risk mitigation and reduction measures implemented
- Clear and unambiguous statements on the level of acceptable risk
- Frameworks, global methodologies and interfaces with other business functions
- Material concerning prevention, i.e., controls and safeguards within the organization
- Risk / event / disaster scenarios derived from analytical steps and risk assessment

The auditor should map the existing set of documentation against the detailed requirements in the BCI standard or other risk management standards utilized. While one audit objective is the verification of completeness, another is to determine the degree of adequacy and plausibility presented by the documents. Risk evaluation and control is sometimes regarded as a separate and

[42] cf. PMBOK or other standardized project management frameworks.

distinct field of research, but is nevertheless an important prerequisite to further BCM activities. Hence, documents should outline the foundation of scenarios and realistic situations derived from perceived risk, consequences, potential losses and the current level of preventative measures existing within the organization. Likewise, the underlying assumptions for risks and their individual evaluation should be tested. In most cases, the risk identification phase will have included several external sources such as historical data or official statistics. The auditor should verify that these sources are both credible and interpreted correctly in the context of the auditee organization's risk evaluation framework.

Area 3: Business Impact Analysis

The Business Impact Analysis (BIA) stage of a BCM process may be a static or a dynamic (repetitive) function within the auditee environment. Documentation should therefore encompass both the methodology used, and the schedule of BIA updates and amendments. The auditor should assess the methodological part of the BIA as well as the implementation over a sufficiently long period of time. In order to ascertain the validity of calculations performed, some assistance by financial auditors or the finance department within the auditee organization is advisable, particularly where documentation about process outages and the corresponding losses are concerned. Depending upon the organization's financial accounting and controlling systems, the underlying calculations for "loss" in a broad sense may be complex.

The BIA represents the central set of documents leading to one side of the strategic BCM question of investment vs. potential loss. The auditor should verify that, notwithstanding the correct and comprehensive calculation of detailed impacts, the BIA process is repeated at an acceptable frequency and – where appropriate – triggered by significant corporate events such as mergers, divestitures, acquisitions and restructurings. This second part of business impact related documentations represents a proper audit trail over a defined period of time which should cover at least one financial year and, preferably, one or more interim BIA runs proving the adequacy of the method used for major changes.

Typical documents requested by the auditor might include outlines and summaries as listed below. In many cases, sampling of detailed documents or drill-downs should be applied to ensure that the central assumptions regarding impact and loss are correct and realistic at the time of audit.

- Definitions and glossaries, alignment with finance and accounting
- Methodologies and calculation templates for losses and impacts
- Summaries of impact and loss calculations for all business processes (drill down or sample where necessary)
- Plausibility of potential impacts, alignment with risk assessment and evaluation (see above)
- Audit trail of BIA activities for at least one financial year
- Audit trail of at least one "ad hoc" BIA done in the course of major changes within the organization

The end result of any BIA should be the prioritization of business functions and processes, thus identifying the "critical" processes and the desired order of recovery and restoration. These priorities are an important part of the documentation presented to senior management for further strategic assessment and, ultimately, the formulation of a BCM strategy. The auditor should verify if and how the detailed BIA results have been converted into suitable summary documents for senior management decision support. Related documents may include, but are not limited to:

- BIA summary reports
- Prioritized business process (function) models
- Lists of business processes that are seen as "critical"
- Materials outlining the order of recovery and restoration for business processes and technical functions
- Decision support matrices and tables
- Summaries of maximum tolerable downtimes (MTD) and maximum tolerable times in alternative operations (MTA)

The main audit objective is to ascertain whether the procedural steps towards preparing summaries and conclusions are complete and adequate. The contents of management presentations – for instance where a highly critical situation presents itself, without any apparent prevention in place – should only be the subject of the auditor's interpretation if serious findings or immediate danger to the auditee organization are evident.

Area 4: BCM Strategy

Following the compilation of the set of documentation for risk evaluation and business impact assessment, the audit should verify that clear and unambiguous strategic objectives have been set for the BCM process, usually expressed by the existence of two or more strategic options. These options should be documented using the correct methodology and a suitable style to allow strategic management decisions, i.e., commit budgets for significant investments. If no BCM strategy exists, or if the strategy itself is insufficiently documented, the auditor should highlight these gaps during the reporting phase. Typical documents supporting a comprehensive BCM strategy include:

- Audit trail for assessment and discussion of strategic alternatives, including viability assessments
- Documents showing a final consolidation of numerous strategic options (usually in the shape of management summaries)
- Documentation indicating business unit and senior management consensus
- Management presentations, requests for clarification, decisions
- Final approval of strategy, management support
- External contracts and integration of services / products
- Links to technical recovery strategies

The outcome of strategic discussions and subsequent decisions should be clearly documented. Furthermore, the auditor is well advised to examine in detail the discussions and methods applied to the "viability" question, namely cost-benefit analyses, references to BIA or risk evaluation results and the conscious integration by senior management of a defined acceptable level of residual risk.

Area 5: Emergency Response and Operations

Emergency response, including incident management, should be seen as a work area adjacent to the overall BCM process. The auditor should obtain a set of documents giving evidence to the fact that an incident or emergency can be controlled by the auditee organization, including the alignment with the larger scope of business continuity. Not all emergency operations documents may be relevant in the context of a BCM audit. In order to assess the readiness for emergency response and its adequate representation within business continuity management, the auditor might request documents such as:

- Lists of potential emergencies and other critical incidents
- Responses envisaged
- Documents outlining the procedures to be followed immediately after an incident
- Cross-referencing with the overall BCM process
- Command structures and emergency organization (e. g. command centers)
- Materials for analyzing past incidents and distilling "lessons learned"
- Historical documentation on past incidents

In addition to the supporting evidence given by the set of documents compiled by the auditor, the adequacy and compliance aspects should be considered. This relates to both legal aspects and other procedural issues that may arise in conjunction with certain types of incidents or emergencies. For instance, the documentation may be comprehensive and integrated with BCM whilst neglecting the requirements set down by public authorities.

Area 6: Detailed Planning

The focus of document requests should be on the planning process rather than plan contents. The auditor should refrain from collecting detailed plans for the business units or functions, as these would not necessarily allow any conclusions as to the preparation, plan formulation and validation. In terms of auditing the detailed planning stage of a BCM process, it is more important to verify if planning is guided by the strategic framework, supported by a suitable organizational model, and validated by appropriate means. The audit objective is to ensure that detailed plans have undergone a reasonable process of drafting, validation and approval rather than criticizing detailed contents. Hence, the set of documents used to support audit findings and observations would include:

- Planning methodology statements ("how-to" documents and other supporting materials)
- BCM organizational charts and relationships with business units
- Interim reporting templates and schedules outlining interaction between BCM staff and business unit personnel
- Validation procedures and assessment templates for plan drafts
- Approvals procedures and maintenance / improvement schedules
- Audit trail for validation and approval of "first versions" of plans
- Links to training and awareness program (see below)

When using the standardized audit program, it is nevertheless possible to extend the audit steps to cover contents and the relative adequacy of plans. However, the auditor should refer to the audit scope and planning statements to determine the appropriate level of detail for assessing detailed plans. Otherwise, collecting extensive documentation and validating it independently may turn out to be a time-consuming effort. It is important to distinguish between two audit objectives which determine the type of documents that the auditor should request:

- Have detailed plans been produced, are they complete, validated and approved using reasonable and adequate procedures?
- Given the set of detailed plans, are they adequate in terms of ensuring business continuity for the auditee organization?

If the second question is to be answered by the auditor, the procedural documents listed above should be complemented by a detailed plan review and assessment, preferably using a prearranged and approved standard audit program.

Area 7: Training and Awareness

Corporate BCM awareness and corresponding training activities are a prerequisite to implementing and improving business continuity. Documentation in this area is often extensive, as the organization is likely to actively improve the level of awareness among its employees. The auditor should not only collect relevant documentation, but verify its intra-organizational impact with regard to BCM exercises and program maintenance. Some awareness programs may appear comprehensive and targeted on the surface, but several prerequisites must be present. The commissioning of corporate business continuity awareness programs should therefore be audited in conjunction with the overall BC framework, particularly the factual recognition of business continuity management in day-to-day business.

Apart from the contents of awareness and training programs and scripts, the compulsory participation of employees – at regular intervals, depending on the function that personnel are required to perform in a BC setting – should be examined in the course of the audit. The factual participation and continuing education about business continuity should be documented, preferably as an audit trail (signed attendance sheets etc.). The documentation on BCM awareness and training might therefore include:

- Internal brochures and other communications encouraging and explaining the BCM process
- Training programs, schedules and manuals
- Timelines, regular training courses and attendance records
- Training contents, alignment with BCM framework and planning
- Etc.

Practical implementations of corporate awareness and training programs have shown that, depending upon corporate culture and the overall training situation, the setup and documentation may vary considerably. The auditor should therefore focus on the overall level of management commitment to this area, rather than follow standardized formats for awareness documents and training schedules.

Area 8: Exercise and Maintenance

Exercising and maintaining the business continuity process should be well documented throughout the auditee organization. The auditor should place particular emphasis on obtaining and scrutinizing the test planning, exercise logs and maintenance schedules, as these indicate whether the current BCM is up to date and adequately reflects the organization's capability to recover from a critical event. Planning, conducting and analyzing the exercise should be supported by a comprehensive set of documents, as should subsequent maintenance. Hence, documents to be requested by the auditor should include:

- Exercise schedules, test plans, preferably for more than one year
- Exercise logs and reports on results, including analyses and "post mortem" documents

- Historical analyses of real incidents or crises and lessons learned
- Regular maintenance schedules
- Ad hoc maintenance logs, e.g., following major organizational changes
- Technical and managerial approval documents
- Software-generated or other audit trails of changes (where appropriate)

The set of documents compiled in the course of the audit should clearly show if and how exercise and maintenance schedules ensure that business continuity planning and management are monitored and updated on a regular basis, and ad hoc as triggered by real incidents or critical events.

Area 9: Public Relations

Public relations and communications as part of crisis management and subsequent activities towards the general public are important components to be verified in the course of the audit. The documents to be reviewed should include:

- General PR strategy, approvals by senior management
- Evidence of PR training for those individuals involved in communicating with the public
- Structured press documentation and predefined statements
- Lists of preferred press and media contacts
- Shadow web sites and similar tools

The auditor should examine the documentation to form an opinion as to whether it reflects the thoughts and intentions of senior management. As management will determine how the entity under review should be presented to the public in the event of a crisis or disaster, it is irrelevant whether the auditor agrees with the factual content of the documents. However, any of the materials reviewed should clearly indicate that exercises or "dry runs" have taken place to validate the PR concept. The auditor should look for cross-references to Area 7 (Training and Awareness) to ensure that all necessary steps have been taken to support the PR concept selected by senior management.

Area 10: Coordination with Public Authorities

In order to determine the level of alignment between the auditee entity and relevant public authorities, some written agreements or understandings should exist. The auditor should also verify that the contents of business continuity plans and management tasks are compliant with regulations or other procedural guidance issued by the local authorities with a legitimate public interest. Hence, documentation should include:

- Minutes, letters of agreement and other evidence showing a clear and unambiguous understanding between the entity under review and public authorities
- Documents stating compliance with requirements and regulations
- Integration plans outlining the interaction between auditee and public authorities in the event of a crisis or disaster.

The standardized audit program contains several audit items that the auditor might use to request documents and other evidence in this work area. An important point to note is the plausibility of the business continuity plans and the corresponding management process in the light of prevailing regulations or preferences expressed by any public authorities involved. The integration of plans or measures supplied by public authorities is particularly important where police forces, fire brigade or other such units are likely to intervene in crisis situations.

Methods

The choice of analytical methods for both interview notes and additional documentation will usually vary in line with the audit plan, the budget and the timeframe envisioned for completing the different audit stages. The material collected rarely lends itself to direct quantification. It is therefore recommended to categorize the findings using a predetermined rating. While this will not necessarily provide any information as to the absolute meaning or severity of a finding, it can serve as a ranking for comparative analysis. It is the auditor's choice whether a verbal or numeric scale is used to denote the opinion on each item. However, over-simplification in terms of categorizing should be avoided. Some typical examples of verbal and numeric schemes are shown below[43]:

Audit Item / Rating	1	2	3	4	5
Item 1					

Audit Item / Rating	n/a	weak	parts	comprehensive	complete
Item 1					

These ratings and scales should be applied to all audit items as given by the audit plan and the program used to cover the ten work areas. It should be noted that, when using verbal descriptions for the ratings, they should be equidistant in terms of their meaning and connotations, to allow the auditor to determine a clear rating for each item. For further analysis, these verbal descriptions will require the assignment of values as an intermediate step.

As a second step, the auditor should refer to the risk level attributed to each audit item, as listed in the program. The risks will normally be categorized in a qualitative manner only, for instance as "high," "medium" or "low." In conjunction with the assessment of the materials gathered beforehand, a weighted rating may be obtained.

In most cases, some cross-referencing between individual audit items will be necessary in order to ascertain the relative significance of a finding. Hence, the use of the standardized audit program is suggested as a means of identifying these interdependencies when assessing individual items. For more complex audit plans, it is recommended that the ratings be subdivided into relevant groups, e.g. completeness, adequacy, plausibility. The resulting table of ratings is outlined below:

Audit Item / Topic	Completeness	Adequacy	Plausibility / Realism
Item 1	4	3	3
Item 2	1	1	1
...	...		

[43] The design of categories and response classes or ratings has been researched extensively. The reader is referred to the body of literature for further details. For the purpose of an aide-memoire, the auditor may safely assume that a simple five-point rating or similar device will be adequate.

Depending upon the overall question to be answered by the audit (*see Chapter 1*), the notions of "completeness," "adequacy" and "realism" may have to be substituted with other wordings, for instance in a due diligence audit situation where the primary aim is different. If the overall audit objective differs from a standard compliance type, the mere fact that all audit items are compliant with standards and regulations would not be regarded as particularly interesting. Hence, tables for mapping the individual ratings should be set up in accordance with the audit scope. The table below shows a typical example that utilizes maturity levels and comparative ratings:

Compliance vs. Maturity

Audit Item	Compl.	Mat. Co. A	Mat. Co. B	Rating	Comment
Item 1	3	3	1	3	use A method
Item 2	5	5	1	5	replace method used at company B with that used within A (= buying in knowledge)
Item 3	1	1	2	1	ad hoc in both companies, not compliant, action req'd
...	...				

Using three fictitious audit items, the table assigns a compliance rating against general standards ("Compl.") and then assesses the level of maturity found in the two companies that form part of the due diligence ("Mat. Co. A," "Mat. Co. B"). The overall rating is a combination of all three assessments, from the point of view of company B who is planning to take over company A. As the principal of the due diligence audit, B is naturally interested in a number of things:

- the areas where both companies have a compliance issue (action is required irrespective of the due diligence)
- the area where company A (the target) can bring additional knowledge and expertise that would come for free if the takeover is effected (particularly audit item 2 which is attractive to B)
- the overall rating that allows an overall assessment of the compliance and maturity aspects

In order to accomplish a uniform rating across the three criteria, relative levels of importance should be assigned to each criterion beforehand, in line with the interests of the due diligence audit. Notwithstanding the above example, the auditor may decide to use similar rating schemes that are better suited to the interests of the company initiating the due diligence.

In a similar manner, other audit objectives should be translated into appropriate rating schemes, using a set of criteria that closely reflect the question to be answered by the overall audit. For typical updates, this may include criteria such as "improved," "unchanged," "deteriorated" and so forth.

However, these ratings may become complex and will inhibit further analysis when the table is incomplete or if some of the ratings are assigned with a degree of uncertainty. The numeric notation may convey an impression of quantified material, and it is often difficult to revisit the reasoning behind a particular rating. Practical experience dictates that this is a likely source of conflict when submitting and discussing the subsequent audit report. If, on the other hand, the auditor should decide to insert references to individual working papers or documents gathered from the auditee, the effort spent on collecting and analyzing the material may exceed the budgetary and time constraints imposed by the audit plan. It is therefore recommended that the auditor focus on completing a fair and impartial opinion of the situation – based upon experience and audit findings – that avoids mechanistic conclusions drawn from ratings and tables.

It is recommended that the standardized audit program be used to obtain a uniform set of ratings and weightings for analytical purposes. However, some adaptations may be necessary to accommodate specific audit objectives.

Analytical Example

The following example shows how a detailed audit area with its items might be rated in accordance with the overall audit objective. Using the Electronic Widgets audit example used in the previous chapters, the compliance dimension is to be considered as well as the due diligence aspect expressed in the audit scope statements (planned merger with German Widgets AG, see above). Hence, two ratings are applied to satisfy both criteria. The compatibility between both companies intending to merge is difficult to ascertain at this stage, given that only one company has been audited. However, the example below provides a clear picture of the situation at Electronic Widgets, thus allowing other auditors to use the results and analyses as a point of reference for their audit steps respectively.

Electronic Widgets
Analysis of Audit Findings
Area 2: Business Impact Assessment

Audit Item	Compl	Adeq	Plaus	Overall Maturity
Business processes identified	4	4	5	4
Process criticality identified	2	1	1	1
3rd party processes analyzed	1	1	1	1
Process interdependencies	3	2	4	3
Interruption insurance	5	3	3	4
MTD´s established	2	1	3	2
MTA´s established	1	1	1	1

The business processes at Electronic Widgets have been identified and formalized to a large degree. Some weaknesses exist with regard to flow charts and other devices which would make the individual process more transparent to an outside observer. However, the realism / plausibility of the process model appears to be adequate, as these business processes have not changed significantly over the past five years. It may be safely assumed that core, support and infrastructure elements are well known within the organization, and that employees have a comprehensive amount of collective experience at their disposal.

Criticality of processes in financial terms has not yet been analyzed properly. Some attempts have been made to distinguish between core business (profit centers) and other business (cost centers), but not with a view to disruptions or damage.

Third party (outsourced) parts of the business have not been covered by any analytical steps or assessments. At the time of audit, no information was available as to their robustness, criticality or potential impact on Electronic Widgets. Service level agreements could not be furnished.

Process interdependencies are well known throughout the organization, but this knowledge is informal and partially restricted to key personnel. As a result, the interdependencies outlined by very senior personnel (>20 years experience in the firm) are very plausible. However, not all areas of business are covered in this way, and no documentation exists that would crystallize the experience of the few knowledgeable individuals.
Business interruption insurance has been contracted for all events deemed relevant by senior management, namely fire, force majeure, supplier failure etc. However, the level of coverage (max. EUR 5,000,000) appears too low, given the average turnover of Electronic Widgets. As a result, both adequacy and plausibility cannot be confirmed for this item.
Maximum tolerable downtimes (MTD) for core processes (a limited set of processes only) have been thought about, and the management has a reasonable, but instinctive notion about the downtimes that might be accepted without permanent damage to the business. However, other processes and cascading effects have never been considered. Most MTA thinking is centered around specific events such as fire or flooding.
Maximum tolerable times in alternative operations (MTA) have never been addressed, as no alternative operations are envisaged. Management has indicated that in the event of a crisis or disaster, one would have to improvise.

[The first part shows overall compliance ratings (simplified) for Electronic Widgets. This is the basis for further analysis in the due diligence context. Note that the EW internal ratings should be grouped and compared with other work areas, as this example only shows Area 2. Depending on the entity under review, some areas should be compared across the Electronic Widgets organization, whereas others may be audited in an isolated manner. The level of detail shown in this example is directly derived from audit scope, the resulting audit plan and on site audit activities performed to complete the tasks set by the audit plan.]

Audit Item	E Widgets	G Widgets	D. D. Rating
Business processes identified	4	2 ?	neg.
Process criticality identified	1	2 ?	pos.
3rd party processes analyzed	1	4	pos.
Process interdependencies	3	3	neutral
Interruption insurance	4	3	neg.
MTDs established	2	1 ?	neg.
MTAs established	1	1 ?	neutral

Given the compliance ratings assigned to the different items within the business impact assessment, a comparison with German Widgets AG is limited, as GWA does not form part of the audit. However, some informed estimates were made on the basis of knowledge that Electronic Widget possesses. The ratings with a question mark should therefore be seen as preliminary.
Business processes at GWA are not well documented, and some weaknesses may exist in their modeling technique. This indicates an overall investment requirement for Electronic Widgets in the event of a merger or takeover.
GWA appears to have developed a measure of process criticality which is based upon core, support and infrastructure processes. However, their criticality is derived from the business model itself and not from potential damage resulting from disruptions. This nevertheless indicates a practice or set of practices which could be assimilated by Electronic Widgets.
As a result of comprehensive SAP outsourcing activities, GWA has carried out an in-depth assessment of third party processes and their significance to the business. The methods applied follow recognized standards and appear to have a high degree of plausibility and realism. The outsourcing assessment has not been extended to smaller processes or partial processes that GWA has outsourced; at the time of audit, the practice seems to be limited to SAP-related activities. For Electronic Widgets, the knowledge acquired would be regarded as a positive element in a takeover or merger.
Process interdependencies are identified and evaluated in a similar manner in both companies. This audit item is therefore neutral in terms of a takeover or merger.
GWA has a limited business interruption insurance policy with high cover but a very limited set of events. Hence, the adequacy rating is low, as there may be many unforeseen events not covered by the insurance policy. It would be advisable to extend the interruption insurance cover.
Maximum tolerable downtimes are not known for GWA. The fact that no documentation exists indicates that

the methods and procedures may not have been defined yet. Within the context of due diligence considerations, this might require additional investment by Electronic Widgets.
Maximum tolerable times in alternative operations have not been defined at GWA. The situation appears similar to the one found within Electronic Widgets.

[The second part compares EW ratings with those of German Widgets AG insofar as the latter are known. Due to the limits of the audit, some uncertainty is inevitable, unless GWA has a data room or other accessible source of information. As a result, the wording in this example reflects the somewhat speculative nature of audit statements.]

Applying the Standardized Program

The standardized audit program in section II contains clusters of audit items as well as suggested risk ratings for each item. All ten work areas are covered, allowing the auditor a uniform application of questions and audit activities across organizational units. Depending on time and effort envisaged for each area, detailed items are decimalized to facilitate planning. The following picture of a typical audit item – in the format of the standardized program – shows how analytical steps may be derived from the standard.

	Risk	Mitigating Controls & Workpaper References
2.3.4 A security audit has been performed to specifically address security-related risks.	High	

Security Concern/Implication
If security as a dominant category of risks has not been addressed in a specific and comprehensive manner, the complex threats and vulnerabilities may not have been envisaged. This may lead to increased exposure of the organization against security breaches.

Recommendation
A security audit should be performed to specifically address the category of risks and threats represented by physical and logical security.

Manual Review Procedure
Obtain and review the security audit program as well as results, findings and reports. Verify that the security audit is comprehensive and in line with recognized security standards or audit standards.

Automated Tool Reference
N/A

Detailed audit activities to be performed:

- Review general audit findings with regard to completeness, relevance and ranking.
- Verify that internal security (both physical and logical) has been assessed.
- Verify that external security (both physical and logical) has been assessed. An attack and penetration test should have been performed. Obtain and check the results.
- Verify that organizational / procedural audit steps have been performed.

Figure 1: Standard frame for audit items according to standardized program format

The decimal header denoting the audit item proper indicates that the question asked belongs to audit area 2 (risk evaluation and control), and to the third sub-group within that area. The sub-group consists of several items, the security item being the fourth question on the list. To the right of the audit item, the suggested significance in terms of risk is given (High). In figure 1, no mitigating controls or general references are available. However, in some cases the auditor may find references to standard works or specific journal articles, audit opinions or other relevant published material. The tabular section below the first part (towards the middle of the frame) shows possible consequences or concerns in conjunction with this audit item. These points may be used by the auditor to explain to the auditee how the absence of a control or the weakness with regard to an audit item may lead to a tangible risk or even danger. Below, a sample recommendation is given for later use in reporting. The right hand column contains "shorthand" descriptions of suggested audit activities and document requests, as well as automated tools to be used where applicable. The auditor should note that both the consequence / recommendation and activities / tools column are indicative and by no means exhaustive.

The bottom part of the standardized frame lists detailed audit activities for the item. These include document review, specific verification and random checks, sampling and suggested interviews with selected individuals. The auditor should direct these detailed activities within the time and budget constraints of the audit; it is unlikely that all points on any of the lists may be covered in the course of a "normal" audit. Most of the detailed steps listed at the bottom of the frame have been tested in practice and been found to be effective in establishing the necessary level of certainty and documentation.

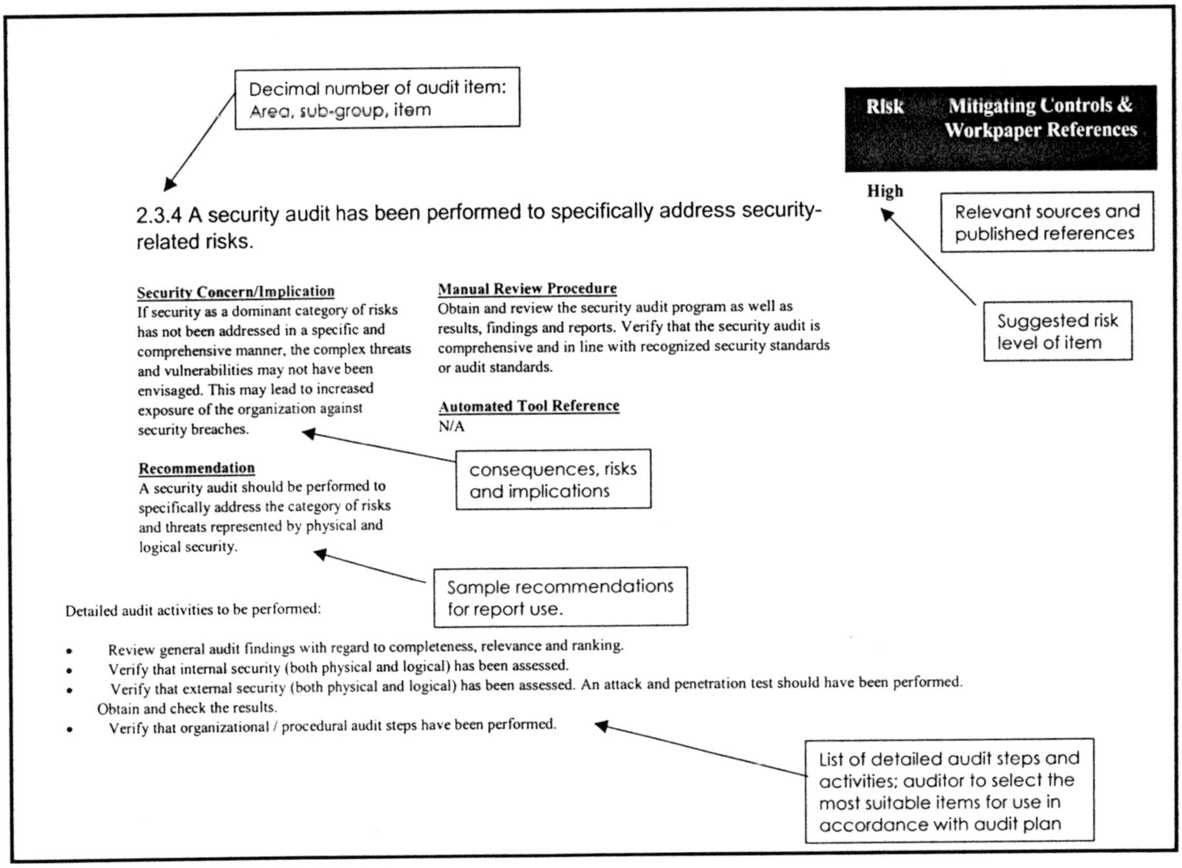

Figure 2: Standard frame with annotations

All steps described in this chapter may be performed using the single frame for each audit item:

- Interviews (see detailed list of audit activities)
- Reference to accepted standards, regulations and other sources (top right hand corner, mitigating controls and workpaper references)
- Risk weightings (top right hand corner)

The ratings for each item will follow from the amount of documentation and other findings. Obviously, the standard frame cannot predetermine any ratings which are an expression of the real situation within the auditee organization. However, *Section II* contains sample rating worksheets for each work area that match the five-point rating scheme suggested above. In order to maximize efficiency when using the standardized program, the following order of activities is recommended:

- Schedule interviews and detailed audit activities according to audit plan, risk level and time or budget constraints
- Drill down where appropriate, using the list of detailed audit activities
- Note findings and cross-reference working papers on the standard frame
- Use secondary sources as indicated in the top right hand corner of the standard frame, and compare with findings for validation purposes
- On the basis of documentation and other material gathered, assign general (compliance) rating for the audit item, preferably using completeness, adequacy and plausibility / realism parameters
- Where appropriate, assign a second rating that relates to the audit scope, e.g. due diligence, annual update etc.
- If so desired, use additional analytical tools such as boundary (zig-zag) diagrams or threshold diagrams to illustrate characteristic patterns and highlight findings.

Despite the standardization attempted by the program, it should be noted that typical lateral conclusions drawn by the auditor are usually based on experience rather than quantitative and qualitative analysis.

Summary

- Compile and summarize interview notes and other incidental material. Perform textual analysis to bring verbal statements in line with the structure of the audit program.
- Systematically identify gaps in the contents of interviews, and follow up as appropriate. Arrange and conduct drill-down interviews to obtain the desired level of information depth.
- Obtain relevant documentation by work area. Compare documentary evidence with interview statements and other sources of information (e.g. walkthrough, physical inspection, sampling).
- Refer to impartial secondary sources, such as literature, standards and regulations to validate the preliminary findings.
- On the basis of the above, form an audit opinion that should reflect both the interest of the audit sponsor and the "yardstick" set externally (legal, regulatory, industry standard).
- Assign risk weightings to individual audit items (cf. the standardized program) to distinguish between critical items and others, e.g. "high," "medium," "low."
- Define criteria for rating the factual findings. One set of criteria should show the level of compliance within the auditee entity, i.e. in terms of completeness, adequacy, and plausibility of BCM-related concepts and measures. Another set of criteria may be needed to express the specific interest of the audit sponsor, e.g. "improved," "unchanged," "deteriorated" for auditing BCM on an annual update basis.
- Group the rated findings in accordance with the audit program. Highlight any unusual findings and risks. Provide detailed descriptions as appropriate to avoid ambiguity and minimize the need for later clarifications.
- Double-check that all planned audit steps have been completed, analyzed and subjected to ratings. This is particularly important for complex, large audit projects.

4

REPORTING GUIDELINES

The success of your work as a BCM auditor depends on clear, concise audit reports that are easily understood by management. Chapter 4 explains how audit reports are structured, written and presented to your stakeholders. In this chapter, you will find samples and templates ranging from small, detailed reports to a large set of reports designed for an international BCM audit.

The third essential component of any BCM audit is the way in which results are communicated to the auditee or other interested parties. In an analogy to financial audit, the design and contents of reports should represent a coherent framework within which the auditor may express their opinions and place the necessary emphasis upon certain findings or circumstances. While financial reporting has come to denote a whole discipline of instruments for compiling or visualizing information, the reporting "universe" for business continuity management is less well defined. Notwithstanding the general assumption that management will read no more than the executive summary pages, it is particularly important in a BCM audit setting to provide an overall structure as well as elaborate information for the different departments or business functions involved in the audit. The decisive factor in determining the success or failure of a reporting system is therefore the degree of detail provided to the individual, the business process owner, and the audited entity as a whole.

By definition, the reports distributed as a result of prior audit activities are the only means of conveying the auditor's opinion. This is less than trivial, given the amount of knowledge that the auditor will inevitably have accumulated over a period of days, weeks or even months. More often than not, this knowledge will prove an impediment to reporting, quite simply because the wealth of data and minute conclusions appears overwhelmingly comprehensive, and condensing the salient facts into adequate reports for different management echelons is a seemingly hopeless task. Whilst working within the constraints of deadlines, budgets and limited resources, the auditor is nevertheless expected to submit a complete set of reports, outlining both factual findings and an answer to the overall issue addressed by the audit (*see chapter 1*). The difficulties increase with the amount of data to be categorized, documented and properly cross-referenced.

Structuring Report Contents

BCM Findings and Audit Statements

It is therefore recommended that a structure be defined for different types of reports and other working papers. The auditor should adopt a "top-down" approach wherever possible, allowing targeted reporting towards different audiences, and standardization of report contents in more complex audit assignments. Conversely, writing the report papers is sometimes easier using the reverse "bottom-up" direction, as this allows compiling single findings and analytical results into clusters which are easier to summarize at the more strategic reporting levels.

Moreover, it is suggested that BCM reports, as a result of their subject matter, require some more elaborate attachments and reference systems than other audit reports. The issues that may arise from the analytical part are complex in themselves, but the report file is required to identify interfaces to more detailed information about the matter at hand. In this sense, BCM reports are perhaps more "academic" in their approach, and the auditor is well advised to envisage an unusual amount of written evidence that should be listed in appropriate appendices or attachments.

Contents broadly fit into one of three categories: descriptive, analytical or normative. The descriptive part of a report simply states facts and figures as determined by the preceding audit steps. While this may appear like stating the obvious, it should be noted that not all auditee personnel may be as familiar with the facts as the auditor, and due to the many areas of work covered, the compilation of an overview of factual statements is often the first step to ensure that all parties involved can find a common basis for discussion.

Having completed the analysis phase, the main benefit introduced by a BCM audit should be to validate the descriptive statements against known BCM standards, weightings and business objectives. Hence, the report should clearly indicate how facts and findings relate to predetermined benchmarks or to an "ideal world" which is usually difficult to achieve but nevertheless demanded by standards and regulations. The auditor should avoid an overly detailed representation of analytical steps and methods in order to focus on the validity and severity of findings. For the auditee, it is more important to know about audit findings and the auditor's opinion on their significance than to study lengthy analytical treatments of questionnaires. The normative part of a report, based upon both the facts and their validation, should indicate clear recommendations and practical steps to be taken within the auditee organization. Recommendations are often translated into projects at the organizational level, and it is not uncommon for auditees to initiate major changes following audit findings.

Practical experience shows that recommended improvements in BCM are often time-consuming and complex projects in a technical and organizational sense. As a result of these considerations, BCM recommendations are often read and subsequently processed by senior management. For instance, a general overhaul of physical security and fire protection will inevitably turn out to be an expensive and lengthy undertaking. It is obvious that such projects cannot proceed without senior management approval which will only be given if the audit finding, its consequences and the corresponding recommendations are expressed in a concise and unambiguous manner. Normative statements should also take into account the overall motive of the audit, namely compliance, due diligence and so forth.

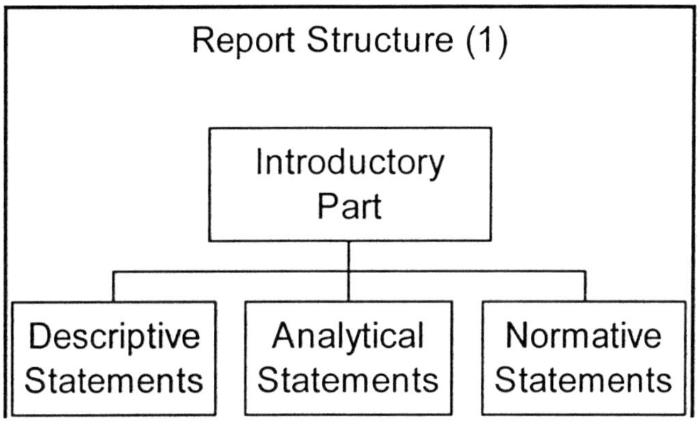

This basic structure complements the organizational dimension in the context of the overall audit plan. In order to adequately address the needs of the auditee organization structure, the audit findings and recommendations should be both general and specific, to distinguish between different areas of work, different business units, and findings relating to parts of the auditee entity or entities.

Structure of Multiple Reports

The framework for these categories of findings is set by the audit plan (*see chapter 1*), thus suggesting a simple system of visualization. This "layered" approach ensures transparency and reiterates the scope of the audit as well as the audit plan. Moreover, it supports the toolbox approach towards reporting suggested here, in that it allows the auditor to build up individual reports to a whole, in a coherent and clear manner. For each of the audited companies or organizations, one layer is added to the reporting "universe." The number of layers is not restricted, but for the sake of clarity, the auditor might like to subdivide the reports into different volumes.

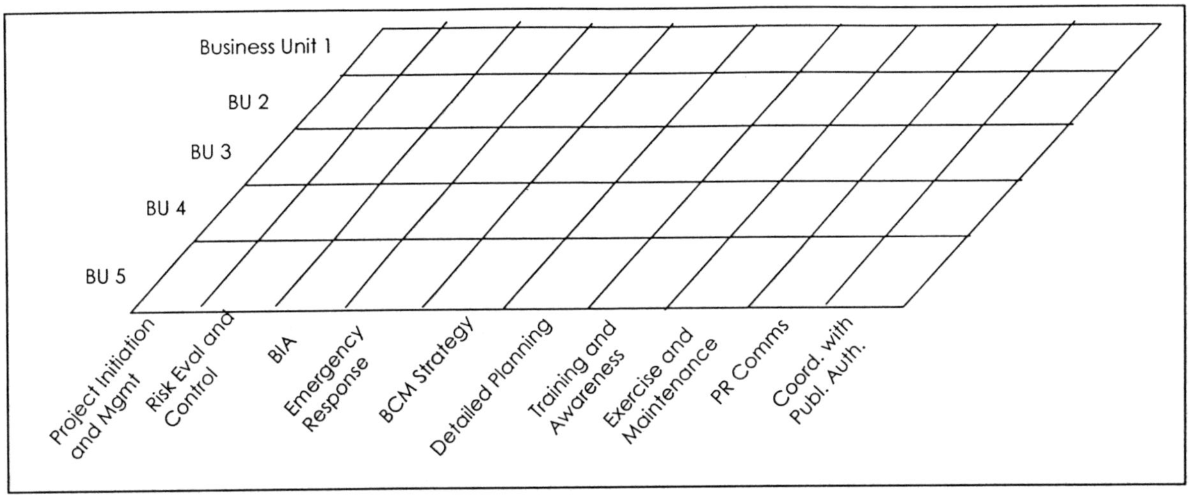

Figure 3a: Visualization Layer for audit report (single)

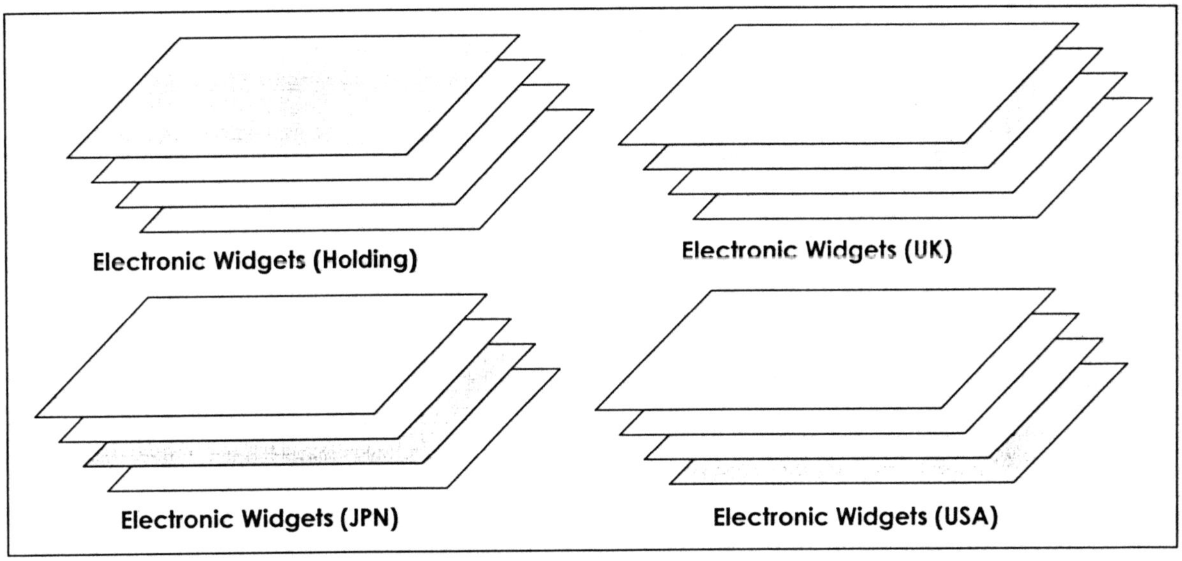

Figure 3b: Visualization Layers for audit report (multiple)

It is recommended that these layers be used throughout the reports to provide guidance to the reader, particularly where complex audits across several organizations are to be reported. The different layers can be built up to complete the picture step by step, and interim reporting may be symbolized by different colors or other suitable devices.

Within each layer, the different work areas are visualized in a similar manner, in accordance with the audit plan. Figure 3c shows how a simple matrix is added to the layer being discussed, indicating to the reader which areas are to be covered and how these fit into the overall picture. Practical experience has shown that these visualization aids will usually help the reader to navigate within and between individual reports. It is therefore suggested that the visual symbols be repeated as part of the page header or footer. The multiple-layer visualization should be used for the report title page to signal the place of the whole report in a suite of similar documents.

A "unifying document" in the shape of a summary report should always be provided. This report does not contain any findings or detailed audit opinions. It is designed to give an overview of the structure and summary findings that the auditor wishes to bring to the attention of senior management. Any details should be cross-referenced, such that the interested reader is able to peruse a detailed report where necessary.

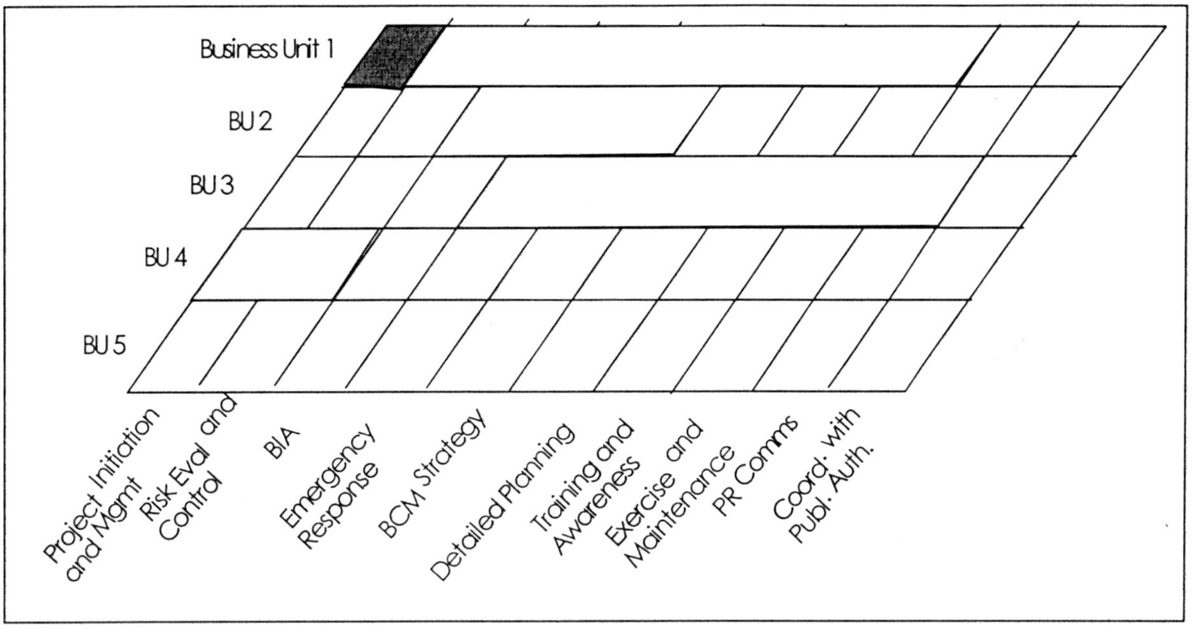

Figure 3c: Visualization layer for audit report, work areas and area being discussed

For the summary report, the audit scope and plan should be included – at least as an outline – to introduce the reader to the BCM work areas and the overall objectives. Methodology, approach, analytical tools used and the reporting structure itself should be explained in a concise manner. The auditor may wish to include some important findings that merit senior management attention, or a brief summary of countries, regions etc. Figure 4 shows how the different documents should be combined to address all relevant reporting levels.

Figure 4: Reporting hierarchy

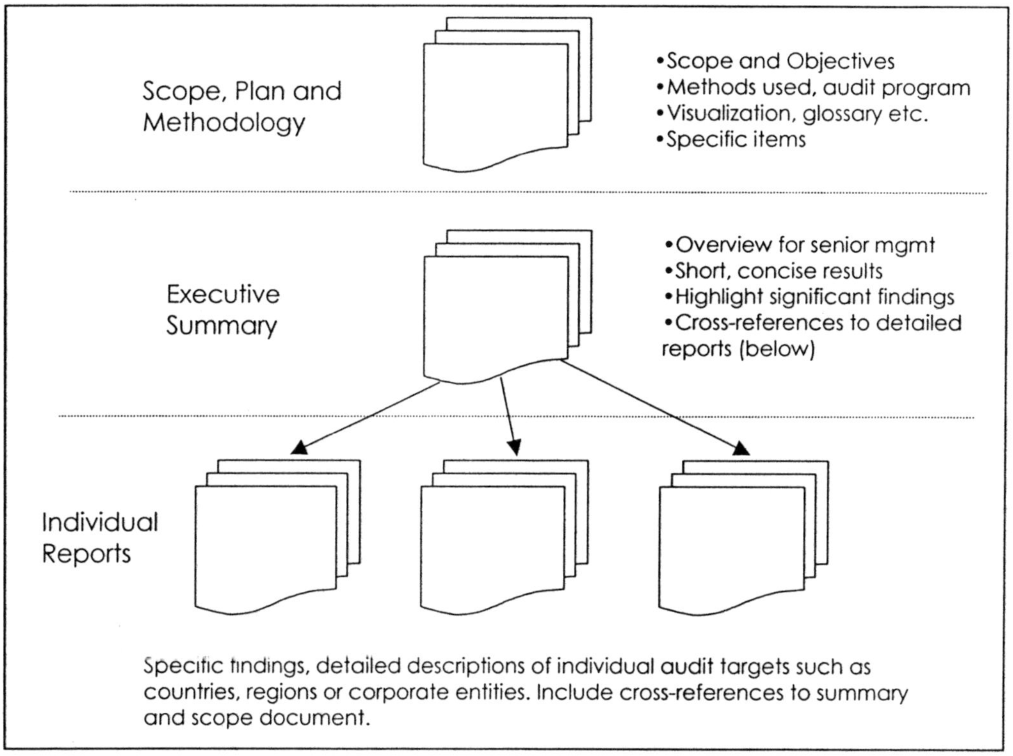

Considering the complexity of audit findings in various business units, and the difficulties presented by the often technical nature of weaknesses, the auditor should include appropriate glossaries, explanations of BCM terms and additional technical references to illustrate specific facts.

Example of Overall Report Structure

The following example illustrates how a set of reports might be prepared and grouped under a uniform structure. Once again the case of Electronic Widgets is used to show the transition from the analysis phase to the initial structuring steps for the multiple reports. Given that the audit plan outlines several countries and work areas to be covered, appropriate references to the audit planning and program documents have been inserted. In the context of this book, the audit report has not been formatted properly, as this would require too much space. It is however recommended that formatting guidelines and issues regarding the auditee organization's corporate identity be adhered to. An example of a full (formatted) report is given in the appendix.

Electronic Widgets
Summary Report Title Page

prepared by <Auditor>
Version 1.0
<Date>

Audit of Business Continuity Management for the Electronic Widgets Group of Companies

Volume I: Reporting System, Structure and Management Summary

Management Summary

As part of the due diligence activities resulting from the proposed merger between Electronic Widgets Group (EWG) and German Widgets AG (GW), an audit of business continuity management (BCM) was performed between <date> and <date>. The objectives were to establish the current state of business continuity within the group, and to obtain an impartial opinion as to the compatibility of BCM between EWG and GW. The Professional Practices of the Business Continuity Institute (BCI) and the Disaster Recovery Institute International (DRII) were applied as a basic standard for verification of compliance and good BCM practice.

The scope and extent of the review covered several countries and work areas within EWG. A detailed description of audit targets and issues addressed is given in the documents "Audit Framework", "Audit Scope" and the individual audit plans issued for the countries and companies respectively. The standard work areas included were:

Area 1: project management and initiation
Area 2: risk evaluation and control
Area 3: business impact assessment
Area 4: emergency response and operations
Area 5: business continuity strategies
Area 6: detailed continuity planning
Area 7: training and awareness
Area 8: exercise and maintenance of business continuity concepts and plans
Area 9: public relations and communications
Area 10: coordination with public authorities

Not all areas were reviewed in all countries; however, the preliminary discussions with senior managers of EWG resulted in the selection of a representative set of audit targets (entities) and areas to be covered. Further areas to be focused on at country or entity level were determined by means of a control self-assessment carried out by local managers.

Findings

Key Issues:

- BCM is heterogeneous. Companies and countries are subdivided into three major categories.
- A majority of countries and companies does not have an adequate level of BCM.
- The absence of BCM has created a number of issues, notably legal problems and official warnings.
- No group-wide BCM frameworks or policies appear to be in existence.

In contrast to the results of the aforementioned self-assessment, business continuity management practices at EWG must be considered heterogeneous and not compliant with recognized standards. With the exception of three countries (Japan, United Kingdom, United States), the general level of BCM and associated planning shows considerable variance. A set of recommendations was developed to indicate potential improvements.

Companies and countries may be subdivided into three major groups. The first group has reached a high level of business continuity awareness and management experience, due to the comparatively long period of evolving plans and concepts. It appears that the early initiatives in these countries were prompted by a comprehensive risk analysis, whose initiation began after major natural disasters and other environmental influences. The second group consists of countries and companies with a limited and uneven BCM capability. While some of the members of this group have comprehensive BCM frameworks in place, the quality and the management process were found to be deficient. Others were found to be operating sophisticated IT-related continuity processes while neglecting other business areas within their corporate structures. The third group encompasses companies and countries where no initiatives have been taken to ensure business continuity and its integration with the business as such. The audit revealed that several larger countries (most notably Germany) had to be assigned to the third group, thus leading to significant continuity-related risks for EWG.

From a work area perspective (see above), most technical areas such as detailed planning and risk evaluation have been addressed in a comprehensive manner, with several exceptions in the second and third group of audit targets. However, several other areas were identified where even the entities in group 1 had not initiated any steps towards compliance with accepted standards and good practices. Areas 9 and 10 were covered in the United States, but only partially in the United Kingdom. No other country has addressed these areas. Area 7 was found to be constrained by local budget considerations, leading to some misunderstandings as to the purpose and meaning of BCM in most countries. Significant weaknesses were identified in Area 4 – crisis and emergency management appear to present difficulties in most countries, as there are no trained personnel to assist in planning and verifying activities in this area.

The overall methodology applied to business continuity at EWG has not been implemented uniformly throughout the group. As a result, many entities have opted for a different approach, often supported by local BCM consultants. While these isolated efforts have produced acceptable results in many companies, it is nevertheless difficult to ascertain whether plans and concepts could be integrated under a unified framework. The audit further revealed that no budgeting, controlling and reporting requirements are in place for business continuity management. It was found that some companies have included BCM in their IT budget, thus restricting the project scope to technical matters. In these cases, local management appeared to be unaware of the fact that BCM had taken place at all.

Several singular risks and findings were deemed important enough to include them in this summary. In two countries, senior management refused to approve BCM initiatives taken by local technical personnel. Three countries experienced disastrous incidents in 1999 and 2000. BCM was perceived as an "unnecessary burden on the budget" in eight countries. Legal action has been taken against one of the companies reviewed, on grounds of not having a BCM and thus having caused material damage to a third party. Official warnings have been served to two companies as a result of not satisfying minimum continuity (disaster and security management) criteria set by local regulations.

Recommendations

Key Action Points:

- There is a need for a group-wide, formalized BCM policy and framework. It should define "best practices" for all group entities.
- In accordance with the framework, a roadmap should be established to outline the way to BCM maturity. The roadmap should be supported by formal project management and a project office.
- Projects should be prioritized to provide further guidance to local management: immediate action, medium-term (financial year), long-term (several years).
- BCM-related funding should be subdivided into balance sheet items (asset investments) and expenditure to provide a clear perspective to group management.
- Internal senior BCM personnel and external consultants should be brought together in task force units to support global deployment of the BCM framework.

The results of the overall audit have been documented in country and company reports. Steps and measures to improve individual weaknesses in business continuity have been recommended. From a group level perspective, the following recommendations address the need for strategic, group-wide initiatives:

Business continuity as a process within companies and entities should be formalized by initiating a group-wide, uniform framework for BCM. While this framework need not be detailed, it should establish clear guidelines and communicate a set of management objectives and expectations to all local management groups throughout EWG. It is recommended that the framework incorporate all industry standards and legal provisions governing business continuity, so as to enable EWG to define corporate "best practices" and minimum requirements.

A comprehensive roadmap should be developed to outline the way towards BCM maturity. Steps, actions and milestones should be included to enable entities within the EWG group to benchmark against known standards. Project planning and management should be formalized and unified to provide guidance to individual companies and management groups. From an audit point of view, it is further suggested that the project management process for BCM be complemented by continuous internal audit. The overall result of managing projects and tracking progress might be seen in terms of continuous improvement and cross-fertilization between companies.

Projects should be grouped according to priorities identified by EWG senior management. It is recommended that a project office be established at group board level to address the following issues:

1) prioritization of projects: immediate action, medium-term initiatives (within one financial year), long-term strategic initiatives (spanning several financial years)
2) liaison with EWG companies and local BCM units: assist in project scoping, planning and project control; cross-fertilize by identifying and communicating "best practices"
3) liaison with Internal Audit: maintain and update an annual audit plan, sponsored by senior management; direct internal audit teams, process audit results and prepare improvement plans

Given this somewhat ambitious vision of a BCM program for EWG, it should be noted that a majority of projects and strategic initiatives may be seen as balance sheet items rather than mere expenditure – for instance, the introduction of group-wide tools or technical improvements at company level may be regarded as long-term investments. It is therefore recommended that the overall project roadmap be included in strategic (group-level) budgeting, in order to distinguish between BCM-related activities that represent expenditure, and those representing balance sheet items (investments).

It is likely that, considering the findings of the audit, EWG will require some external (consulting) support to accomplish the goals set by a group-wide BCM strategy. Current examples indicate that the level of external resources utilized in some countries has been adequate, and outcomes have been satisfactory. In contrast to audit activities, the subsequent implementation of these recommendations will be complex and difficult to control or monitor at company level without deploying experienced, impartial external resources whose direct reporting relationship with the group BCM project office will avoid local misunderstandings. A task force approach is suggested, using both senior project managers from internal departments and external consultants working in mixed deployment teams.

Audit Structure and Reporting System

The audit of business continuity planning and management was structured in accordance with the scope and planning statement (enclosed). Individual audit findings and recommendations are described in detail in the following documents:

Document Type	Contents
Company Audit Report	country-by-country audit results, findings and local management summaries.
Compliance Report (Due Diligence)	where EWG and GW entities exist in one country or location. Given that there might be a potential merger in these locations, a comparative audit result has been prepared
Overview Report	comparative report of all countries and entities, containing detailed analysis and assessment of BCM
Management Summary	this document

The reporting system is therefore three-tiered. A numbering system for individual report was used to ensure quick referencing and handling of reports. All reports have been submitted on a CD-ROM in an Internet Explorer (HTML) readable format (enclosed). The following numbering conventions were used:

A_AUSTRIA_12-31-01	country Audit report (detailed) with date of audit (last day)
C_AUSTRIA_12-31-01	compliance Due Diligence report (detailed) with date of audit (last day)
A_OVERVIEW	overview Audit report (detailed)
C_OVERVIEW	overview Due Diligence report (detailed)

Additional documentation was referenced accordingly, using the above report name as a prefix and subsequent numberings, such as:

A_AUSTRIA_12-31-01_Area1_SampleDocumentNameAndPurpose
C_JAPAN_10-31-01_Area8_SampleDocumentNameAndPurpose
...

This ensures that all working papers and materials collected in the course of the audit have been assigned to reports in a clear and unambiguous manner. Rather than using a numbering scheme, it was seen as more transparent to make full use of long filenames and standard data formats where appropriate.

All documents gathered and thus classified were converted to the following data formats:

- Standard EWG documents (electronic): Portable Document Format (PDF) with digital signatures
- Standard audit working papers (electronic): Microsoft Word format
- Paper copies and originals (scanned): Tagged Image File Format (TIF) with digital signature, additional PDF copies with digital signature for reasons of space and e-mail sizes
- Photographs: negatives, scanned TIF, converted PDF for reasons of space and e-mail sizes
- Worksheets, databases and other complex files: original format such as Excel, Oracle and others, secondary PGP-encrypted copies for later comparison
- Correspondence (paper): filed originals
- Correspondence (e-mail): filed printouts, e-mail archive files in some cases

Programs and tools required to use the CD-ROM containing the sum of all audit-related material:

- Adobe Acrobat Reader v3.0 or higher (freeware)
- Microsoft Word 97 or higher (alternative packages with a converter may be used)
- Microsoft Excel 97 or higher (alternative packages with a converter may be used)
- ODBC recommended for some database files
- PGP v7.0 or higher if decryption of reference copies is required
- Any picture editor or graphics manipulation package for scanned files and photographs (PDF available, picture editor required for originals only)
- Microsoft Internet Explorer or other web-enabled viewer

[next table: List of all audit reports prepared with file names]

The full report would contain a listing of all audit files and reports compiled. This has not been included in the example for reasons of space and clarity. The example also shows that with complex audit reporting, paper copies and files may be unwieldy. Using a CD-ROM may be advisable when dealing with large amounts of data and diversified file formats.

Miscellaneous Reporting Issues

Technical References

In contrast to other audit fields, particularly finance, it is likely that many findings will require footnotes to provide additional evidence or references to books or articles. A bibliography is therefore recommended for an overview of relevant publications. While it may be argued that the managerial reader is unlikely to read such technical references, they are nevertheless important for the more operational levels of the corporate hierarchy, for instance in facilities management and information technology. Hence, technical findings should be explained at an appropriate level of details, using attachments and appendices where necessary. When referring to published or unpublished material in this way, the auditor might wish to include copies of articles and papers with the audit report. Unlike an academic publication, the audit report should not require further research on part of the reader, although copyright restrictions may limit the amount of material that the auditor can copy and provide as a matter of courtesy.

In many cases, technical references may appear cumbersome and somewhat exaggerated – at first sight. However, the auditor is often required to provide evidence, or prove the technical points that constitute an audit finding. While the financial auditor can quite easily refer to standard works known to most financial experts, the BCM specialist is often forced to compile supporting evidence from many sources published in different periodicals and languages. This in turn requires that an audit report be much more specific and detailed in its referencing system and bibliography.

Management Presentations

Many audit engagements require final presentations to management which should be based upon the audit findings and reports. In most cases, these presentations will be given to senior management or the board of directors. Consequently, they differ significantly from the detailed reporting schedule outlined above. It is recommended that any presentation be short, concise and focused on the main objectives pursued both by business continuity management within the auditee organization and the audit itself. Given that presentations are usually prepared using an appropriate graphics package, the majority of information is likely to be given orally, rather than in writing. Experience has shown that elaborate presentations containing numerous slides may be counter-productive. BCM, more so than other audit topics, requires particular attention to brevity and a concise style of presenting. Otherwise, findings or recommendations may take a long time to explain or visualize. The auditor should limit presentation items in line with the management summary (see example above) to ensure that the audience is addressed at an appropriate technical level, and in a concise manner. The following rules of thumb, although fairly generalized in nature, nevertheless apply to presenting BCM findings and recommendations to a senior management group:

- it takes at least three minutes to present one slide. Each question from the audience will prolong this time by one to two minutes.
- it takes thirty seconds for the reader to understand a complex visual slide
- it takes thirty seconds for the audience to read a text slide with many bullet points
- a slide with more than four bullet points (textual) appears crowded
- if the presentation contains more than seven major findings, they should be sub-grouped. Otherwise, the audience will not remember all of them
- when using tabular representations of numbers or graphs, add thirty to sixty seconds time for

understanding, and up to two minutes for explaining what the visual means

Using these simple rules, the auditor may calculate overall presentation time by counting slides and estimating the overall duration – with and without questions from the audience. It is further suggested that longer presentations be structured to provide natural breaks or audience participation. This is particularly important when the contents of the presentation address several topics:

- when a question is formed in the mind of a person listening to the presentation, this question will remain "active" for 30 seconds (short-term memory), for up to four hours (medium-term memory) or indefinitely (long-term memory). Some BCM questions relate to complex facts and visuals. It is advisable to increase the number of feedback loops by asking for questions from the audience after each "complicated" slide – this will cover the questions in short-term memory.
- Not all questions are necessarily noted down by participants. The auditor may prefer to continue without interruption. It is recommended to use a flip chart to physically note down the questions for later discussion, and to reassure the audience that all questions will be answered.
- Rather than asking for management input and opinions after the presentation, the auditor may provide "breaks" by asking for statements from the audience during the presentations. These may be useful when completing one aspect of BCM, such as risk management, prior to proceeding to the next one (business impact assessment). This type of audience participation, as well as asking for action points set by management, will enable the auditor / presenter to reinforce the rapport with the audience, and to give positive feedback while progressing towards the conclusions ("as Mr. X has pointed out earlier, this slide shows that BIA activities have not been addressed in countries A, B and C...").

The auditor, in presenting to management groups, should take into consideration that BCM – in most cases – will be an unfamiliar topic to the majority of the audience. While some knowledgeable individuals may be present, others will be confronting the issues of business continuity for the first time. It is therefore very important that all participants be made comfortable with the subject, and that ample discussion time be provided. Ideally, senior managers should themselves formulate the conclusions and action points arising from the presentation, rather than being deprived of their prerogative to make decisions. Despite the fact that the auditor is presenting findings and recommendations, the latter should be formulated in a manner that invites managerial reactions and decisions, rather than pre-empting these decisions.

Applying the Standardized Audit Program

The standardized audit program provides sample risks, their immediate consequences and sample recommendations:

Fig. 5: Annotated single frame, standardized audit program

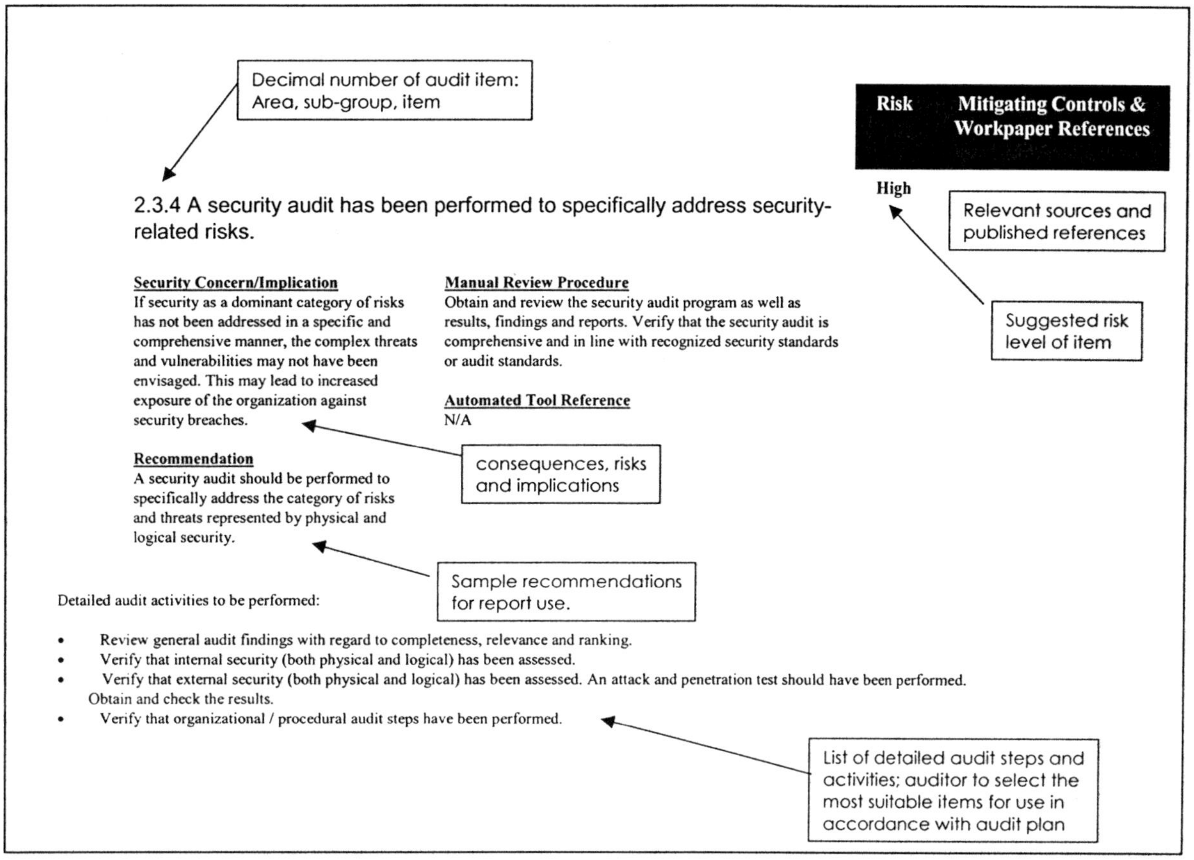

Following the analysis phase of the BCM audit, the auditor will have completed a list of findings, their relative severity, and possible consequences for the entity under review. However, the immediate output of the analysis phase is a rating, and perhaps a tabular representation of the issue. For reporting purposes, the ratings and analytical results should be transformed into an appropriate text output, taking into account both the facts and their assessment.

On the basis of the detailed audit activities to be performed, a clear picture of the situation should exist. The sample risk section entitled "security concerns / implications" denotes a short summary that is indicative of the risk intrinsic to the audit item. Reporting these implications may require some additional thought on part of the auditor, as findings or generalized risks may need adjustments to reflect the specific situation. The sample risks are therefore no more than a useful hint which the auditor may use as a starting point for reporting. Consequences of a missing or inadequately implemented BCM component, for instance the absence of an authorization mechanism for funding and cash expenses during a crisis, must be seen in context: the auditor should decide whether the entity under review is likely to suffer damage from this particular fact. In a small, relatively unstructured company, this may not be the case, as the managing director will react swiftly and simply nominate several individuals to act as signatories to cash outlays. In larger firms, the opposite may happen – confusion as to authorizations and the individuals empowered to actually spend money may severely impede the functioning of the finance department and other departments having to rely upon timely payments to creditors and suppliers. Hence, the auditor should decide on the basis of the organizational context rather than prefabricated statements, and each audit report may contain significantly different wording.

The sample recommendations section should be used in a similar manner. The standardized audit program provides no more than a first indicator as to the general nature of recommendations. The auditor should beware of copying the wording in the program. All recommendations should be formulated with a view to addressing the background and specific situation of the auditee organization. Statements such as "an authorization mechanism for cash expenses in crisis situations should be introduced" should be elaborated by the auditor, for instance by referring to the current authorization guidelines and their possible transformation to a crisis situation: "the current practice of requiring at least two signatures for any cash transaction should be reviewed and simplified for crisis management purposes. It is recommended that any payment be authorized by the leader of the crisis management team and any director or officer present at the time."

Both findings and recommendations should be calibrated in terms of their verbiage. When using the standardized program, findings, possible consequences and recommended actions should be uniform in terms of the words used:

- **severity of finding**: define typical levels of using "risk," "significant risk," "dangerous risk," etc. to mirror the ratings previously assigned (analysis phase). Specifically mention the risk ratings where appropriate.
- **possible consequences**: describe consequences at a uniform level of detail – consequences elaborated on in the report denote a higher risk. Distinguish between financial and other implications.
- **recommendations**: define typical levels of urgency: "recommend," "strongly recommend;" use of words such as "immediately," "shortly," and so forth. Apply these wordings uniformly throughout the audit reports.

Calibration is quite straightforward if the standardized ratings for the audit program are used. Standard wordings may be mapped against the numerical values, as indicated below:

General risk rating of audit item	Assessment rating of entity under review	Finding	Recommendation
High	1	significant risk immediate danger direct threat ...	strongly recommend immediately without delay ...
High	5	no apparent risk danger unlikely no threat in evidence ...	suggest currently satisfactory no action required ...
Medium	1	existing risk indirectly leading to danger of indirect / derived threat ...	highly recommended within an appropriate timeframe medium-term ...
Medium	5	no foreseeable risk no foreseeable danger very limited potential threat ...	Recommend as part of long-term improvement initiative long-term ...
Low	1	remote risk danger unlikely ...	suggest less significant later ...
Low	5	...	...

The table outlines some possible wordings that may be assigned to different numerical values. Clearly, a "1" in conjunction with a generally high risk represented by the audit item would lead to fairly decisive language in an audit report, whereas a "5" (mature) in conjunction with a generally low audit risk for the item would not merit any further description in the report.

Summary

- Structure reports in terms of factual, analytical, and normative statements (i.e. findings, possible implications, recommendations)
- The reporting system should reflect the audit scope and planning statements: use an overall reporting framework that is adequate to the task
- Multiple reports should be categorized and grouped to allow aggregation to higher (senior management) levels
- The different levels of reporting should consider the target groups: management summary – senior management, individual audit report – local personnel
- Parts or the whole of the reporting system may be on paper, on CD-ROM or in any other suitable format that is most convenient to the audit sponsor
- Report items should reflect the results of the analysis phase. "Calibrated" wordings should be used to avoid discrepancies in verbiage
- Standardized risks, implications and recommendations are of an indicative nature only. Refer to the situation and the organizational background when reporting
- Closely control the referencing system and any cross-references in reports
- Prepare management presentations using the rules of thumb identified. Avoid long presentations
- Explain BCM-related technicalities for the audience, but do not pre-empt decisions
- Interact with the audience during management presentations. Provide feedback loops and elicit action items and decisions
- Calculate the length and duration of management presentations to determine the level of granularity that is appropriate to the occasion.

SECTION II

STANDARDIZED AUDIT PROGRAM

Section 2 is a standardized audit program divided into work areas. You will find detailed audit questions covering all aspects of business continuity management. In the course of your BCM audit, you can use parts or the whole of the standardized questions for your audit plan and program. The standardized audit program is designed to give you additional information on risk ratings, recognized standards and additional materials that you may use to understand each item, as well as to communicate it to audit teams or the auditee organization.

For each item within the standardized audit program, the legal, regulatory and technical background is explained in detail. Detailed audit steps have been included for each question to give you indications as to the time and effort required during the audit. Suggested standard wordings for findings and recommendations have also been included.

1 Project Initiation and Management			
1.1 Scope, Objectives and Format		**Risk**	**Workpaper References**
1.1.1 A formal enterprise-wide business continuity plan exists.		**High**	
Security Concern/Implication By not having a formal enterprise-wide business continuity plan, there is an increased risk that the organization will not be able to recover from a disruption in a timely and effective manner. In some countries and / or industries, both business continuity and risk management may be mandatory. Risks not mitigated by BCM may have implications re: going concern. **Recommendation**	**Manual Review Procedure** Query management as to whether they have a formal business continuity plan. Review the plan to determine whether it encompasses the entire corporate entity or entities reviewed. Observe third party relationships where applicable. **Automated Tool Reference** N/A		

Detailed audit activities to be performed:

- Describe and explain the terminology used and refer to client terminology where appropriate. Establish a glossary where necessary. Keep in mind that some firms may prefer terms such as "contingency planning," "crisis management," "disaster recovery," etc.
- Refer to any existing business process models for further information
- Assess the legal and regulatory background for the business unit(s) under review - any legal requirements for BCM should be addressed immediately and formulated as a finding. Examples include ISO/IEC 17799 (BS 7799), Kontroll- und Transparenzgesetz (KonTraG, Business Control and Transparency Act), Trading Room Recovery Guidelines of Institut Monetaire du Luxembourg (IML) etc., see also Chapter 11 in this section
- Plans may be informal in the sense that several people have agreed procedures and call trees among themselves. While this may be an indicator of client awareness, it does not satisfy the criteria of a formal BCM. Likewise, the DRP / BCM approach - when restricted to IT - is not adequate in the sense of this question.

			Risk	Workpaper References
	1.1.2 The plan covers the entire set of business processes and corporate entities.		High	ARIS Manual, VISIO Manual
	<u>Security Concern/Implication</u> If business processes and / or corporate entities, such as subsidiaries or joint ventures, are not included in the business plan, it may be difficult or impossible to recover these units in the event of a disaster or major incident. <u>Recommendation</u>	<u>Manual Review Procedure</u> Review the organizational structure as well as general and specific business process models. Determine the boundaries of the audit universe and define entities or processes which may be beyond the control of the audit sponsor (3rd parties). <u>Automated Tool Reference</u> IDS ARIS Toolset, Visio, Flowcharting packages		business process models and charts supplied by the organization

Detailed audit activities to be performed:

- BCM should address all business processes, or at least a significant proportion of critical business processes which may, in some firms, be called "business functions." Insular business continuity planning and/or management activities do not satisfy the criteria for formal and comprehensive BC.
- Using the auditee's documentation, identify all corporate entities that are wholly or partially owned or controlled. In some countries there may be joint venture setups that are mandated by local law but constitute a subsidiary relationship.
- Obtain high-level business process documentation for all entities. Employ automated tools where appropriate.

		Risk	Workpaper References
1.1.3 Introductory information, such as scope, objective, and assumptions is included in the business continuity plan.		Low	
Security Concern/Implication If the scope, objectives and assumptions are not included in the business continuity plan, there could be misinterpretations about what the plan encompasses and what is the objective of the document. This could lead to improper use of the plan document. **Recommendation** We recommend including the scope, objective, and assumptions in the business continuity plan to inform all personnel of what is contained within the document.	**Manual Review Procedure** Review the business continuity plan to ensure that the scope, objective and assumptions are included in the documentation. **Automated Tool Reference** N/A		Introductory BCP documents, scope statement, objectives lists

Detailed audit activities to be performed:

- Review completeness of information as above
- Assess adequacy of assumptions, scope (beyond IT) and objectives.
- Test employee awareness, senior management awareness

			Risk	Workpaper References
	1.1.4 The business continuity planning and management process is sponsored by senior management.		High	
	Security Concern/Implication If the business continuity project is not sponsored by senior management, its effectiveness will be limited. Planning and managing continuity may not meet the requirements. Budget, resource allocation and set timeframes cannot be enforced. **Recommendation**	**Manual Review Procedure** Review documentation relating to project initiation and confirm senior management involvement. Check approval, charter, and other sponsorship evidence. **Automated Tool Reference**		Formal memos and other communications showing senior management involvement

Detailed audit steps to be performed:

- If possible, conduct personal interviews with members of senior management (specifically the sponsor). Test awareness and up-to-date knowledge that the interviewees possess.
- Obtain and peruse documents indicating sponsorship, e.g. minutes of kick-off meetings, formal memoranda, project charters.
- Verify that the person or committee sponsoring the business continuity project is still active and / or employed by the company. If management changes have occurred, confirm that ongoing commitment to the project exists.

		Risk	Workpaper References
1.1.5 The business continuity plan is governed by a formal project management lifecycle or methodology.		High	Project Management Body of Knowledge 2000 (PMBOK)[44] Kerzner (1998)[45]
Security Concern/Implication If the business continuity plan is not conceptually secured by a formal project management lifecycle or methodology, the actual outcomes of the project may not meet defined objectives. Delays, budget overruns and compromise on quality may be evident. **Recommendation** The business continuity planning and management effort should follow a pre-defined and formalised project management lifecycle, including a recognised methodology. These should adhere to industry standards wherever possible.	**Manual Review Procedure** Review project documentation from initiation to completion and post-project analysis. **Automated Tool Reference**		

Detailed audit steps to be performed:

- Obtain typical documents such as statement of work, project charter, work breakdown structure. Assess the adequacy of the planning phase of the business continuity project. Take into account the actual outcome(s) and compare these against the original plan.
- Analyse how the project developed, and assess the controlling method(s) applied. Identify critical steps and milestones. Check whether any deviations from the original project plan were addressed in a comprehensive and decisive manner.
- Compare the actual status of business continuity planning and management with the original plan. Highlight areas where the quality and performance of the BC process (as observed) is below the standards originally defined.

---

[44] available on www.pmi.org and other project management web sites, or directly from the Project Management Institute.

[45] This is recommended as one of the standard summary works on project management and its knowledge areas. See bibliography for details.

1.2 Organizational BCM Integration		Risk	Workpaper References
1.2.1 The business continuity co-ordinator reports to or is sponsored by senior management. Alternatively, a senior manager has assumed responsibility for business continuity management.		**High**	Project stage: PMBOK ™ and other project mgmt. literature. Operating stage: consult literature on organizational risk management
Security Concern/Implication If BCM is not sponsored or managed directly by senior management, the required level of resources or effort may not be available. This may compromise BCM effectiveness and the company's ability to react to business interruptions. **Recommendation** It is recommended that senior management assume ownership of the BCM process, either personally or by assigning the task to a direct report. BCM should be clearly identified as a full-time, critical business process.	**Manual Review Procedure** Review organization charts of client organization. Identify BCM-related positions. Assess matrix organization (if any) and its implications for BCM. Conduct interviews with BCM personnel, senior mgmt. **Automated Tool Reference** N/A		client organization charts and BCM team structure if documented

Detailed audit activities to be performed:

- Document the roles and responsibilities within BCM departments. If functions are assigned to individuals, check whether these BCM responsibilities are part-time, full-time or institutionalised (as staff positions).
- Verify if senior management has recognised BCM as a critical process. Collect correspondence and / or other documentation.
- Perform detailed audit of how responsibility for BCM was assigned, including job descriptions, performance appraisal procedures etc. It is vital to obtain evidence in this sensitive area. BCM should be clearly integrated and recognised by senior mgmt., not just assigned to a person on a part-time basis. The results of this audit step are one of the most important parts of the program.

			Risk	Mitigating Controls & Workpaper References
	1.2.2 BCM Resources have been assigned according to business needs. The position(s) / department(s) are adequately staffed in view of required effort and size of organization.		High	
	Security Concern/Implication If functional assignments and headcount do not match the requirements in terms of effort and size of organization, the planning and management stages of BCM may not be performed correctly and comprehensively. As a result, general BCM capability may not be sufficient in the event of a disaster. **Recommendation** It is recommended that all positions and/or departments are staffed in line with organizational requirements.	**Manual Review Procedure** Review general organization charts. Compare BCM headcount and matrix relationships. **Automated Tool Reference**		as above industry benchmarking data if available

Detailed audit activities to be performed:

- Compare total number of employees with BCM headcount. Take into account the level of complexity inherent in the (known) process model. Assess adequacy of BCM headcount.
- Derive ratios for no. of employees: no. of BCM staff and benchmark against known industry sector average (where applicable).

1.3 Financial Planning and BCM Budget		Risk	Workpaper References
1.3.1 The BCM budget is a separate and distinct budget item.		Medium	
Security Concern/Implication If BCM is not assigned a separate budget, parts of the process may not be funded adequately due to budgetary competition. As a result, business continuity may not be ensured in the event of a disaster. **Recommendation** In our opinion, it is necessary to assign a separate BCM budget whose elements can be reviewed as a distinct budget item.	**Manual Review Procedure** Review budget data. **Automated Tool Reference** Financial accounting tools, MS Excel etc.		general financial budget BCM budget

Detailed audit activities to be performed:

- Review overall budget and identify BCM / BCM-related items.
- Collect evidence by saving copies of relevant files, alternatively take printouts.
- Physically obtain approvals sheets and other relevant documentation that shows how budget planning and budget approval have been achieved for BCM / BCM-related items.

			Risk	Workpaper References
	1.3.2 BCM does not form a part of IT budget spending. Financial responsibility for IT-DRP and BCM does not reside solely with the IT manager.		**High**	
	Security Concern/Implication If BCM is defined as a part of IT spending, there is a significant risk of underfeeding. As a result, there is a high probability that BCM will be dysfunctional in the event of a disaster. **Recommendation** The budget for IT-DRP and BCM should not form part of the overall IT budget.	**Manual Review Procedure** Review IT budget. Check IT-DRP / BCM related items. **Automated Tool Reference** Financial accounting, MS Excel etc.		general financial budget IT budget BCM budget

Detailed audit activities to be performed:

- Obtain copies of all IT budget sheets or business plans.
- Review IT-DRP / BCM related spending and budget items.
- Verify that IT-DRP / BCM items include only those funds that are required for IT as a business process. Budget items for other business processes and their plans should not be part of the IT budget.
- Where other business process IT-DRP items are included, collect evidence and documented explanations. This includes correspondence.

	Risk	Workpaper References
1.3.3 The BCM budget follows BIA findings. It is in line with the original BCM strategy approved by senior management. Sufficient funds have been allocated to planning, testing and maintaining BCM.	High	
Security Concern/Implication If the total budgets for BCM are too low or not in line with the original agreements (BIA, Strategy), the planning process lacks credibility and may be severely impeded in the event of a disaster. **Recommendation** All budgets should be realigned regularly and clearly reflect senior management's decisions. Where insufficient funds have been allocated, the budget should be reviewed. **Manual Review Procedure** Refer to BIA, strategy documentation. Assess viability of actual budget. Identify trends in budgeting (decreasing ??). Document senior management decisions. **Automated Tool Reference**		BIA results documentation and individual calculations BCM budget

Detailed audit activities:

- Review all BIA results and critical business process plans. Compare with actual budgets assigned.
- Benchmark budgets against industry average and / or other independent indicators. Realistically assess the possibility of underfunding and consequences given that one or more incident scenarios occur.
- Document all senior management decisions and the reasoning behind the BCM budget. Where significant underfunding exists, identify overt and hidden reasons.
- Check whether BCM has been reprioritized due to other considerations. Document this.

Overview Chapter 1 Audit Items

Number	Preliminary Rating Assigned by Auditor (from 1 to 5, 1 = absent or inadequate, 5 = comprehensive and adequate), and auditor´s comment	Risk
1.1.1		High
1.1.2		High
1.1.4		High
1.1.5		High
1.2.1		High
1.2.2		High
1.3.2		High
1.3.3		High
1.3.1		Medium
1.1.3		Low

<table>
<tr><td colspan="2"><h2>2. Risk Management and Evaluation</h2></td><td colspan="2"></td></tr>
<tr><td colspan="2"><h3>2.1 Risk Identification, Loss Potentials, Vulnerabilities</h3></td><td>Risk</td><td>Workpaper References</td></tr>
<tr><td colspan="2">2.1.1 The BCM documentation contains clear indications as to the potential losses and vulnerabilities.</td><td>High</td><td></td></tr>
<tr><td><u>Security Concern/Implication</u>
If risks, potential losses and vulnerabilities have not been identified, the BCM may be dysfunctional.

<u>Recommendation</u>
A formal risk identification and classification scheme should be included as part of the BCM methodology. Vulnerabilities should be assessed based upon credible information sources and a realistic expected probability and impact.</td><td><u>Manual Review Procedure</u>
Obtain documentation relating to risks and vulnerabilities. Check classification, impact and probability calculations. Verify if the list of risks and vulnerabilities is realistic.

<u>Automated Tool Reference</u>
N/A</td><td></td><td>List of risks and other documents related to risk analysis</td></tr>
</table>

Detailed audit activities to be performed:

- Verify the list(s) of internal and external threats identified by the auditee organization. Check if these conform with accepted methods of risk classification and risk identification. Consult experts where appropriate.
- Verify the probabilities assigned to each of the risks identified. Obtain documentation for those risks with a low probability and a high impact.
- Check the cost-benefit analysis associated with risks, impacts and any controls that may have been put in place to mitigate the risks.

			Risk	Workpaper References
	2.1.2 An adequate level of preventative controls is in place to ensure that exposure to known risks and threats is minimized.		High	Known standards, e.g. BS 7799 for security, ISO standards for facilities.
	Security Concern/Implication If the set of preventive controls is limited or insufficiently detailed, known risks or threats may endanger the organization. This may increase the amount of potential loss in the event of a crisis or disaster. **Recommendation** An adequate set of preventive controls should exist within the auditee organization. Controls should address the risks, threats and vulnerabilities identified.	**Manual Review Procedure** Obtain and review the documentation relating to preventive controls designed for risk mitigation. Perform in-depth checks where appropriate, for instance in the field of security. Refer to recognized standards as indicated by the work area covered, e.g. BS 7799 **Automated Tool Reference**		ISO 17799 other ISO standards as applicable for country standards refer to chapter 11 of this section

Detailed audit activities to be performed:

- Perform comprehensive examinations of physical facilities and their protection against the risks and threats identified as relevant to the organization. This should include access control, asset management, personnel, logical protection of information and any other specific security considerations. Consult experts where appropriate.
- Assess the "perimeter" of the organization, i.e. its boundaries in a physical and logical sense. Include neighbors or other external factors influencing the protection level of the organization.
- Assess insurance and other additional means of safeguarding the organization against losses. Obtain and verify relevant documentation. Check alignment of arrangements with the risk and vulnerability assessment carried out by the organization.
- Perform IT recovery audit as appropriate. Refer to in-depth works on IT disaster recovery and related subjects for further information. Experts should be brought in in most cases to ensure adequate technical knowledge is applied to the IT side of the audit.
- Obtain and review any documentation relating to procedures and regulations governing the protection measures and preventive controls. Include external personnel, service level agreements and other important factors.

2.2 Risk Analysis Methodologies and Tools		Risk	Workpaper References
2.2.1 Analytical methods and tools used for risk analysis are comprehensive and adequate.		High	
Security Concern/Implication If the methods and results of the organization's risk analysis are flawed, the resulting risk ratings may be inaccurate. This may result in increased exposure to losses. **Recommendation** A comprehensive and adequate set of methods and tools for risk analysis should be in place. It should include quantitative and qualitative elements to ensure adequate modelling and assessment of risks identified.	**Manual Review Procedure** Review the toolbox of quantitative, qualitative and automated risk analysis methods used throughout the organization. Highlight areas of uncertainty and identify potential errors. Verify that the application of the methods identified provides an adequate picture of the risks to the organization. **Automated Tool Reference** Risk analysis and management tools used by the auditee.		

Detailed audit activities to be performed:

- Review quantitative methods used and assess their adequacy, based on the practical implications and the theoretical background. Not all risk analysis tools may be functional within the auditee organization, depending on the quality and density of representative statistical data. Include expert advice and refer to the literature as appropriate.
- Establish the corresponding methods of qualitative risk analysis. Check if these complement the quantitative toolset. Ensure that the residual uncertainty intrinsic to quantitative results is compensated by appropriate qualitative (judgement) methods that may be based on past experience and other "soft" factors. Establish the amount of residual risk and uncertainty that remains despite the complete quantitative / qualitative risk analysis.

		Risk	Workpaper References
2.2.2 Risk analyses are performed in a homogeneous and orderly manner, as well as at regular intervals.		High	
Security Concern/Implication If risk analyses are not performed frequently and in a uniform manner throughout the organization, unknown risks may exist. This in turn may lead to increased exposure and unknown vulnerabilities that may endanger the business. **Recommendation** The risk analysis process throughout the organization should be performed in a homogeneous, orderly manner. Regular intervals should be set for updating the results of the risk analysis and assessment.	**Manual Review Procedure** Review the documentation of previous risk analysis runs. Check that all organizational units are analysed on a regular basis. Compare results and highlight differences and discrepancies. **Automated Tool Reference** N/A		

Detailed audit activities to be performed:

- Obtain the model of quantitative risk analysis. Compare the model with recognized methodologies usually applied throughout the industry. Highlight any differences and document reasons for "non-standard" methods utilized. Verify, using expert help if needed, that the risk analysis methods are comprehensive and represent the state of the art.
- Obtain and review the data gathered for risk analysis purposes. Check if data density and historical data are sufficient to allow meaningful quantification. Identify any spurious or inaccurate use of quantitative methods on small or very small datasets. Highlight any questionable conclusions drawn from non-representative data. Determine confidence levels and find out how much the organization relies on "guesstimates" rather than proper quantitative analysis.
- Compare risk analysis results and reports from different organizational units, using sampling techniques where appropriate. Highlight any divergence from the overall model that may have occurred in any of the selected units or areas. Identify reasons for incomplete or inaccurate use of the methods.
- Repeat the above steps for additional qualitative risk analysis techniques.
- Verify that risk analyses are performed regularly and frequently. Highlight any out-of-date results or organizational units that have not performed an update in the last financial year. Verify that risk analysis results are updated when major changes (e.g. acquisitions, mergers etc.) occur.

		Risk	Workpaper References
2.2.3 An adequate process for ongoing information gathering is in place.		High	
<u>**Security Concern/Implication**</u> If information gathering is heterogeneous and not formalized, the basis for data analysis and other risk analysis activities may not be adequate. As a result, the analytical process may be invalidated. <u>**Recommendation**</u> Information gathering should be formalized throughout the organization. The process should cover all data and information necessary to provide a clear picture that may be used for subsequent analysis.	<u>**Manual Review Procedure**</u> Review procedures and methods for data and information gathering. Check if the results of this process are suitable for further analysis. <u>**Automated Tool Reference**</u> N/A		Questionnaires interview schedules regular meetings BCM committees

Detailed audit activities to be performed:

- Obtain process flows and other documentation related to information gathering about risks, threats and vulnerabilities. Check whether the information gathering approach is used uniformly throughout the organization.
- Review forms, questionnaires, meeting schedules and other procedural material that supports information gathering. Verify that regular meetings take place to update and review information. Check if other instruments are being used for frequent information update, e.g. questionnaires or e-mail surveys.

2.3 Risk Evaluation and Control		Risk	Workpaper References
2.3.1 The business continuity management process contains crisis and disaster scenarios.		High	
Security Concern/Implication If no scenarios have been envisaged, the organization may be inadequately prepared for a sudden critical situation. Without scenarios, testing and updating individual business continuity plans may be difficult. Test results and suggested improvements may not be realistic. **Recommendation** Crisis and disaster scenario of varying severity should be included in the BCM documentation to provide realistic background to testing, maintenance and planning activities.	**Manual Review Procedure** Review scenario descriptions and verify that these reflect the risks, threats and vulnerabilities previously identified. **Automated Tool Reference** N/A		

Detailed audit activities to be performed:

- Review all scenarios that have been proposed by different managers and employees of the auditee organization. Check that the number of scenarios selected is representative, realistic and not confusing in terms of numbers of scenarios.
- Verify that scenarios are "closed," i.e. describe the whole of the organization or the relevant parts for the selected level of severity.
- Check that all scenarios make a worst case assumption regarding direct impact, indirect impact, seasonality etc.
- Verify that scenarios represent varying levels of severity, ranging from smaller events to large impact events. These levels should be in line with the amount of residual risk consciously accepted by senior management. The number of "unknowns" should be minimized.
- Verify that there is at least one scenario aimed at dealing with unforeseen events as they occur.

	Risk	Workpaper References
2.3.2 All risks have been properly evaluated in terms of relevance to the organization, and classified accordingly.	High	
Security Concern/Implication If risks have not been evaluated and classified, their relevance to the organization may not be clear. This may lead to confusion and planning against irrelevant risks. **Recommendation** All risks identified should be evaluated against set criteria, and classified in a simple and unambiguous manner. The classification should take into account the needs and priorities of the business.	**Manual Review Procedure** Review risk classification scheme, evaluation reports and all related communications. Highlight risks that may not have been evaluated correctly. Identify the criteria as to relevance for the organization and its core business activities. **Automated Tool Reference** N/A	risk evaluation reports memos and other correspondence on risk reports

Detailed audit activities to be performed:

- Review documentation relating to individual risks and their evaluation against pre-defined criteria. Include the risk classification scheme and verify that the process of classification is in line with managerial priorities. Where risks of a certain kind may be grouped to reflect specific threats or vulnerabilities (e.g. security), verify that both evaluation and classification are reasonably homogeneous.
- Verify that no personal or other interest has been included in the risk evaluation. Individual fears, concerns and risk perceptions stemming from past experience should not be the basis of an impartial evaluation and classification.

		Risk	Workpaper References
2.3.3 Potential impacts on factors essential for conducting business have been evaluated.		Medium	
Security Concern/Implication If risks, threats and vulnerabilities have not been assessed with regard to their relevance to critical business processes and resources, the magnitude of damage or disruption cannot be estimated. This may increase the potential exposure in the event of a disaster. **Recommendation** All impacts on business processes and resources should be considered when evaluating risk. This need not follow the pattern of a business impact analysis.	**Manual Review Procedure** Verify that the overall impact – not necessarily quantified at this stage – of risks has been assessed. **Automated Tool Reference** N/A		

Detailed audit activities to be performed:

- Refer to the organization's process model to find out whether a mapping of risks against business processes and resources has been done.
- Verify that the listed risks have been evaluated for each process / resource.
- Check whether typical support and infrastructure resources within the organization (IT, facilities, transportation etc.) have been included in the evaluation.

		Risk	Workpaper References
2.3.4 A security audit has been performed to specifically address security-related risks.		Medium	Literature on security audit
Security Concern/Implication If security as a dominant category of risks has not been addressed in a specific and comprehensive manner, the complex threats and vulnerabilities may not have been envisaged. This may lead to increased exposure of the organization against security breaches. **Recommendation** A security audit should be performed to specifically address the category of risks and threats represented by physical and logical security.	**Manual Review Procedure** Obtain and review the security audit program as well as results, findings and reports. Verify that the security audit is comprehensive and in line with recognized security standards or audit standards. **Automated Tool Reference** N/A		

Detailed audit activities to be performed:

- Review general audit findings with regard to completeness, relevance and ranking.
- Verify that internal security (both physical and logical) has been assessed.
- Verify that external security (both physical and logical) has been assessed. An attack and penetration test should have been performed. Obtain and check the results.
- Verify that organizational / procedural audit steps have been performed.

Overview Chapter 2 Audit Items

Number	Preliminary Rating Assigned by Auditor (from 1 to 5, 1 = absent or inadequate, 5 = comprehensive and adequate), and auditor's comment	Risk
2.1.1		High
2.1.2		High
2.2.1		High
2.2.2		High
2.2.3		High
2.3.1		High
2.3.2		High
2.3.3		Medium
2.3.4		Medium

		Risk	Workpaper References
3 Business Impact Analysis Activities			
3.1 A comprehensive business impact analysis has been performed.		High	
Security Concern/Implication If a business impact analysis is not performed, there is no accurate measurement of which business processes are the most critical, what the minimum recovery timeframe is, what resources are needed for recovery, or what the operational and financial impacts to the organization would be if a disruption were to occur. **Recommendation** We recommend conducting a comprehensive business impact analysis.	**Manual Review Procedure** Query management as to whether a comprehensive business impact analysis has been performed. Review documentation for the business impact analysis to ensure that it was complete. **Automated Tool Reference** N/A		

Detailed audit activities to be performed:

- Check method of actuarial or other damage assessment. Indicate method (payback, flat cost, internal rate of return, net present value etc.) of calculation.
- Using samples, verify the BIA calculations.
- Check BIA against general process model (completeness).
- Verify that BIA is not event-driven. BIA should address process failure, not list of risks.
- Verify underlying scenarios, if any.

When using software, refer to chapter 12 of this section.

		Risk	Workpaper References
3.2 A list of prioritized business processes exists.		Medium	
Security Concern/Implication If a list of prioritised business processes does not exist, it is possible that non-critical business processes will be restored before the more critical ones. This will slow down the recovery time of critical business processes. **Recommendation** In order to define critical business processes, it is strongly recommended to document all processes in a formal manner.	**Manual Review Procedure** Review the business continuity plan to determine whether all business processes have been prioritised and documented within the plan. **Automated Tool Reference** N/A		

Detailed audit activities to be performed:

- Examine criteria for prioritisation (across organizational units, across functional departments). Ensure that the original instructions for identification and weighting of "critical" processes have been understood and applied homogeneously across all organizational units. To this end, it may be necessary to review initial training materials and to interview business unit managers.
- Verify if criticality (in terms of time and money) is applied as a unified criterion throughout the organization. Check process ratings in detail.
- Follow BPR / business analysis steps to determine if core, support, and infrastructural processes have been defined in an appropriate manner. The scope of each process and or business function should be evenly distributed, i.e. no massive undefined processes vs. too detailed sub-sub-processes.

	Risk	Workpaper References
3.3 All vendors, suppliers, and third-party companies that are relied upon have a business continuity plan.	Medium	
Security Concern/Implication If vendors, suppliers, and other third-party companies do not have reliable business continuity plans, a disaster affecting one of these companies could also affect all organizations relying on that company. It is possible that internal plans may be compromised if a key supplier is unable to respond to a local disaster (cascading effects) **Recommendation** All vendors, suppliers and other third parties having a business relationship with the company should have reliable business continuity plans. **Manual Review Procedure** Review the business continuity plan and document all vendors, suppliers, and third-party companies that are relied upon. Query management as to whether they are aware if these third parties have business continuity plans in place. **Automated Tool Reference** N/A		

Detailed audit activities to be performed:

- Restrict activities to confirming the EXISTENCE of plans for third parties. For more detailed results, apply this document recursively to the supplier in question.
- Bear in mind that some third parties may not approve of an audit. You should seek client / senior management backing prior to auditing third parties.

		Risk	Workpaper References
3.4 An adequate level of business interruption insurance is established.		Medium	
Security Concern/Implication Without the proper level of business interruption insurance, the costs of recovering the business after a disaster could be prohibitive. **Recommendation** We recommend that business interruption insurance is considered as part of a BCM strategy, to lessen any costs of a disaster.	**Manual Review Procedure** Query management as to whether they have business interruption insurance. Review the policy to ensure that they have adequate coverage. **Automated Tool Reference** N/A		

Detailed audit activities to be performed:

- Obtain policies and other documentation for business interruption insurance
- Examine related policies (fire, theft, other)
- Verify that all insurance policies mitigating against particular risks form a comprehensive framework. Consult insurance experts where appropriate.
- For some countries: verify that mandatory (statutory) insurance has been arranged and confirmed. Example Germany: statutory fire insurance, statutory employee health and safety, etc.
- Where business interruption or related policies are absent, check in detail how the client proposes to cover the risks in an alternative way.

		Risk	Workpaper References
3.5 Business process interdependencies are defined.		High	
Security Concern/Implication If business process interdependencies are not clearly defined, it is possible that a critical business process may not be properly recovered because of a reliance on another business process. This could hinder the recovery of the critical business process. **Recommendation** We recommend that all business process interdependencies are defined.	**Manual Review Procedure** Review the business continuity plan to ensure that all interdependencies are defined. **Automated Tool Reference** N/A		

Detailed audit activities to be performed:

- Apply standard process modelling techniques to confirm interdependencies (ERM, SERM etc.).
- Clearly state possible cascading effects (time, damage) and their consequences in BIA terms.

		Risk	Workpaper References
3.6 Maximum tolerable downtimes (MTDs) are established on the basis of financial and operational impacts of a disruption to normal business operations.		High	
Security Concern/Implication Without MTDs based on financial and operational impacts, it is difficult to determine which business processes are the most critical and, therefore, which of the business processes should be recovered first. **Recommendation** We recommend that maximum tolerable downtimes be calculated on the basis of financial and operational impacts of a disruption to normal business operations.	**Manual Review Procedure** Query management as to how the MTDs were determined. Ask if financial and operations impacts were considered in their determination. **Automated Tool Reference** N/A		

Detailed audit activities to be performed:

- Check if clear distinction exists between MTD and MTA calculations.
- Verify damage over time (business impact) calculations and how MTDs are derived from these.
- Check in detail that all MTDs are based upon impact rather than personal perception of process importance. Overestimating importance, or justification of process existence by means of quoting a zero MTD should not be present in the calculations and results.

	Risk	Workpaper References
3.7 Maximum times in alternative operations (MTAs) for all business processes are defined and documented.	High	
Security Concern/Implication If MTAs are not determined and documented, it will not be clear which business processes need to be restored first. If the critical business processes are not restored within the MTA, it could lead to a substantial loss of business. **Recommendation** We recommend that MTAs are determined for all business processes and that they are documented in the plan. **Manual Review Procedure** Review the business continuity plan to ensure MTAs for all business processes are defined and documented. **Automated Tool Reference** N/A		

Detailed audit activities to be performed:

- Verify that a clear distinction exists between MTD and MTA (see above)
- Assess all assumptions for MTA calculation, i.e. resource requirements, relocation, decentralized work, use of temporary employees etc.
- Assess backlog growth (as a result of alternative ops) and its implications for MTA calculation.
- If there is a disproportionate discrepancy between MTA and MTD (short MTD, long MTA), question assumptions and resource requirements quoted for recovery.

Overview Chapter 3 Audit Items

Number	Preliminary Rating Assigned by Auditor (from 1 to 5, 1 = absent or inadequate, 5 = comprehensive and adequate), and auditor's comment	Risk
3.1		Hig
3.5		Hig
3.6		Hig
3.7		Hig
3.2		Medium
3.3		Medium
3.4		Medium

4 Emergency Response and Operations				
4.1 Command and Control			Risk	Workpaper References
4.1.1 A central location for command and control purposes (emergency operations center, EOC) has been defined in the business continuity plan			High	Literature on the ICS (Incident Command System Auf der Heide (1989)[46]
Security Concern/Implication If a location where key senior management personnel are to assemble following a disaster is not clearly identified in the business continuity plan, there is a potential for confusion which could make the recovery effort more time consuming. **Recommendation** It is recommended that an emergency operations center be defined and described in a comprehensive manner, in order to ensure centralized command and control in the event of a disaster.	**Manual Review Procedure** Review the business continuity plan to ensure that a emergency operations center has been included. **Automated Tool Reference**			

Detailed audit activities to be performed:

- Review the BC plans and related documentation to ascertain whether an emergency operations center in the proper sense of the word exists: the location should be at a sufficient distance from the primary facility, clearly identified and described as a center for command and control purposes, and adequately equipped to perform these functions.
- Verify that multiple locations have been envisaged for command center purposes. In addition to meeting rooms etc., alternative premises must have been contracted for in the event of unavailability of the primary command center.

[46] see bibliography for details on this work

		Risk	Workpaper References
4.1.2 The EOC description adequately reflects the needs of senior management and operational crisis / incident management.		Medium	
Security Concern/Implication If management needs and the specific circumstances of operational crisis or incident management are not recognized and used as a basis for command center design and implementation, the EOC may not be adequately equipped to serve its purpose. This in turn creates the risk of deficient command and control structures in the event of a disruption or disaster. **Recommendation** The description of the EOC should reflect senior management needs and requirements defined by operational crisis and incident managers appointed by the organization.	**Manual Review Procedure** Obtain a comprehensive description of the EOC structure, its location and the corresponding lists of equipment. Compare the documentation with management needs and the requirements formulated by operational crisis and incident managers. **Automated Tool Reference** N/A		EOC documents and charts Equipment lists Location data and documentation

Detailed audit activities to be performed:

- Verify that senior management needs and crisis / incident management needs have been clearly defined and communicated.
- Interview key individuals to ascertain the adequacy of actual vs. desired command center design. Highlight discrepancies.
- Note any vague or unclear statements about management needs and check that the EOC design is flexible enough to accommodate unexpected requests.

		Risk	Workpaper References
4.1.3 Procedures for opening, operating and closing the EOC have been included in the plan.		High	
Security Concern/Implication If no clear procedures are established for opening, operating and closing the EOC, the command and control structure may be impaired. **Recommendation** Opening, operating and closing the EOC should be clearly defined and by means of appropriate procedures.	**Manual Review Procedure** Review all procedures relating to the emergency operations center (EOC) and verify that adequate opening, operating and closing instructions have been given. **Automated Tool Reference** N/A		

Detailed audit activities to be performed:

- Review activation and opening instructions. Focus on authorization, typical trigger actions and alignment with other business continuity plans. The EOC should be the set up immediately after the declaration of a crisis or disaster, authorized by senior management and the BC coordinator. Highlight any weaknesses and ambiguities that may exist in terms of opening and activating the EOC.
- Review rostering and team schedules for the EOC. 24/7 operations should be defined for all EOC activities. Team schedules should be in line with the overall organizational structure and personnel requirements. Key individuals should be scheduled in a way that allows adequate rest periods.
- Obtain and review all operational procedures for the EOC. Verify that all BC plans and related activities are included in the management task lists of the EOC.
- Examine all security instructions for the EOC. Security should be equivalent to the level of security normally applied to sensitive business operations.
- Review closing and de-escalation procedures for the EOC. Focus on authorizations and alignment with recovery / restoration activities in other business continuity plans.

		Risk	Workpaper References
4.1.4 A system for documentation, logging and securing of relevant evidence has been defined for the EOC.		High	
Security Concern/Implication If documentation for EOC actions is limited or insufficient, there may be misunderstandings with regard to later analysis. As a result, claims and other disputes may arise which could lead to corporate and personal liability issues. **Recommendation** All actions taken by the EOC and its staff should be closely monitored, logged and documented. Logging procedures should ensure that all evidence is secured properly and maintained in an appropriate format for later analysis.	**Manual Review Procedure** Review all procedures, forms and logs for the EOC. Highlight weaknesses and areas where evidence is not documented comprehensively. **Automated Tool Reference** N/A		EOC set of forms Procedural documents for EOC

Detailed audit activities to be performed:

- Examine and understand the structure of logs and other documentation. Preferably, these should follow recognized standards, e.g. the ICS. Check for completeness, ease of use and simplicity in terms of applying documents and logs in crisis situations and under time pressure.
- Take random samples of logs and other documents from exercises or real crisis situations. Verify that all documentation requirements have been followed. Highlight any weaknesses.

	Risk	Workpaper References
4.1.5 Arrangements for medical care, food and drink supplies, showering/washing facilities, sleeping/resting facilities, etc. have been made for the EOC and these arrangements are documented in the business continuity plan.	Low	
Security Concern/Implication If arrangements are not made for medical care, food and drink supplies, showering/washing facilities, sleeping/resting facilities, etc., employees may have difficulty working for the extended hours which may be necessary during a recovery, thus slowing the recovery effort. **Recommendation** We recommend that arrangements for medical care, food and drink supplies, showering/washing facilities, sleeping/resting facilities, etc. are made for the alternate facility to ensure the comfort and safety of employees, and that these arrangements are documented in the business continuity plan. **Manual Review Procedure** Query management as to whether arrangements for medical care, food and drink supplies, showering/washing facilities, sleeping/resting facilities, etc. have been made for the alternate facility. Review the business continuity plan to ensure that these arrangements are documented within. **Automated Tool Reference** N/A		

Detailed audit activities to be performed:

- Obtain and review the lists of provisions, materials and other resources for the EOC. Verify that these supplies are sufficient for the purpose of operating the EOC for the defined period of time.
- Review contracts, local delivery arrangements and other means of securing sufficient supplies and required services. Compare these with arrangements that would be in force under normal circumstances. Highlight any weaknesses that might impede EOC operations, or lead to shortages.

4.2 Response Steps		Risk	Workpaper References
4.2.1 Detailed procedures regarding how to activate emergency response are documented.		Medium	
Security Concern/Implication If detailed activation procedures are not documented in the business continuity plan, it is possible for personnel to use alternate activation procedures, which could lead to the plan not being activated in a timely manner. **Recommendation** We recommend that detailed procedures regarding how to activate the business continuity plan are documented to ensure that if a disaster were to occur the plan could be activated immediately.	**Manual Review Procedure** Review the business continuity plan to determine if activation procedures are included and well-documented. **Automated Tool Reference** N/A		

Detailed audit activities to be performed:

- Assess activation authority, levels of authority and chain of command. Determine whether this organizational construct is viable in a crisis situation.
- Review activation steps (action lists)
- Assess efficiency of activation process
- Verify inclusion of authorities in activation procedure
- Verify adequacy of escalation (pre-activation)
- Locate the plan or plans
- Assess accessibility and clarity of instructions, including number of copies, degree of decentralized access to plans, knowledge of employees
- Are there any signs / blackboard notices / other short indicators explaining activation?
- Is there a central telephone number for activation purposes?
- Activation instructions should be clear and concise, resembling a "red button" to be pushed in the event of a crisis.

It is important to distinguish between activation procedures that are officially mandated (fire, first aid, police, etc.) and those pointing towards the internal business continuity organization. While it is certainly necessary to follow general procedures for alerting fire brigade / police / other authorities, a simultaneous activation of internal plans is highly desirable. Short notices and other devices used for employee information should outline the situation and appropriate reaction in no more than 5 concise bullet points, preferably in many locations throughout the company's premises.

		Risk	Workpaper References
4.2.2 Protection of Personnel has been included in the business continuity and emergency response plans.		High	standard works, e.g. Auf der Heide (1989).
Security Concern/Implication If steps for protecting personnel are inadequate or missing, increased exposure and danger to personnel may result. This will raise liability issues and ethical concerns for the organization and its senior managers. **Recommendation** Personnel protection being the first and foremost duty, all procedures and steps for protecting employees and the general public should be documented and strictly enforced in case of an emergency. Statutory provisions must be followed.	**Manual Review Procedure** Review in detail the steps and procedures for the protection of corporate personnel and the general public. Highlight any weaknesses (even small ones) and questionable practices. Ensure that all reasonable precautions have been taken to protect human life and health. **Automated Tool Reference** N/A		

Detailed audit activities to be performed:

- Refer to statutory and other regulatory requirements and provisions to obtain a clear picture of what must be done – as a minimum – by the auditee organization. Consult legal and other experts where appropriate to ensure that the regulatory background is clear and unambiguous. Compare these minimum requirements against the steps and measures outlined in the organization's business continuity and emergency response plans.
- Validate in detail the steps taken to protect corporate personnel, contractors and other individuals who have a direct relationship with the auditee organization. Specifically review all actions envisaged and check for compliance with laws and regulations.
- Validate in detail the steps taken to protect the general public or other individuals who may be on the premises or nearby. Make sure that the business continuity / emergency response plans include actions to protect these people.
- Verify first-aid and other medical steps as well as triage mechanisms. Refer to similar BCM documentations and accepted standards to understand the requirements. Consult experts where appropriate.

		Risk	Workpaper References
4.2.3 Containment measures have been included in the plan. Methods for damage and effect assessment have been defined.		Medium	
Security Concern/Implication If containment does not begin immediately after the critical event or disaster has been declared, the overall damage may increase. If damage assessment activities are not initiated immediately after the critical event or disaster has been declared, damage may be difficult to substantiate, and later insurance claims may be at risk. **Recommendation** Procedures for incident containment and damage assessment should be included in the emergency response part of the overall business continuity plan. Both containment and damage assessment should start immediately after the critical event or disaster has been declared.	**Manual Review Procedure** Review the procedural documentation relating to incident containment and damage assessment. Focus on teams, expertise and detailed steps to be taken immediately after the incident has been declared a critical event or disaster. Highlight any weaknesses and delays. **Automated Tool Reference** N/A		

Detailed audit activities to be performed:

- Obtain and review the detailed procedures for containing an incident. Verify that any containment steps include appropriate escalation methods to ensure timely declaration and involvement of senior management.
- Examine in detail how damage assessment is defined. Validate the methods and calculations used to estimate damage under conditions of uncertainty. Check whether all occurrences of material damage are properly assessed and backed up by appropriate evidence (e.g. photography).
- Verify that containment and damage assessment are done simultaneously.

		Risk	Workpaper References
4.2.4 Emergency response procedures have clear and transparent links to other areas of recovery and restoration.		High	
Security Concern/Implication If response efforts are not integrated with other work areas of the business continuity concept, misunderstandings may occur which lead to confusion, delays and increased damage. **Recommendation** Clear links should be established between the immediate emergency response and other areas of business continuity.	**Manual Review Procedure** Review all cross-references between the emergency response plan and other areas of the BCM concept. Highlight any discrepancies and ambiguities. **Automated Tool Reference** N/A		

Detailed audit activities to be performed:

- Check that the emergency response part of the BCM concept has clear links to the other work areas (refer to the other parts of this audit program), specifically coordination with public authorities (*chapter 10*), detailed business continuity plans (*chapter 6*) and public relations (*chapter 9*).

		Risk	Workpaper References
4.2.5 Salvage and initiation of restoration procedures have been included in the business continuity and emergency response plans.		High	
Security Concern/Implication If salvage procedures are not included in the business continuity plan, increased damage may occur. If initial restoration procedures are not triggered by the immediate emergency response plan, recovery and restoration may be delayed, resulting in higher levels of damage. **Recommendation** Salvage procedures should be included in the plan, as well as initiation of recovery and restoration.	**Manual Review Procedure** Review the detailed steps for salvaging equipment and other corporate resources. Verify that trigger conditions have been included which initiate recovery and restoration. **Automated Tool Reference** N/A		

Detailed audit activities to be performed:

- Review procedures for completeness checks with regard to equipment, facilities and specific items listed as salvageable
- Verify that salvage procedures are aligned with other emergency instructions governing evacuation, personnel safety etc. Salvage should be clearly prioritised so as to prevent dangerous or risky attempts at salvaging equipment.
- Check whether the changeover between immediate salvage of equipment and assets and the subsequent restoration has been described in detail, including financial and other trigger conditions. Highlight any ambiguities that exist in this area.

		Risk	Workpaper References
4.2.6 Procedures for expense authorization during a recovery are explained in the business continuity plan.		Medium	
Security Concern/Implication If procedures for expense authorization during a recovery are not explained, there is a possibility of confusion regarding how to purchase necessary equipment or other items during the recovery. If it is not clear how to pay for these items, they might not be able to be obtained, thus slowing the recovery. **Recommendation** We recommend that procedures for expense authorization during a recovery are explained in the business continuity plan so that all recovery-related expenses can be properly handled.	**Manual Review Procedure** Review the business continuity plan to determine if procedures for expense authorization during a recovery are explained. **Automated Tool Reference** N/A		

Detailed audit activities to be performed:

- Verify completeness of expense authorization mechanisms (all plans, all business units)
- Check adequacy of expense limits - consider crisis or recovery situation and greater spans of control usually assigned to individuals. The internal control system should accommodate the changed situation, i.e., provide audit trails, etc.
- Obtain and review a list of individuals with signatory powers – both under normal operations and in a crisis situation. Compare the two matrices of expense authorization and assess their adequacy in terms of accelerated decision-making in crisis situations or emergencies.

Overview Chapter 4 Audit Items

Number	Preliminary Rating Assigned by Auditor (from 1 to 5, 1 = absent or inadequate, 5 = comprehensive and adequate), and auditor's comment	Risk
4.1.1		High
4.1.3		High
4.1.4		High
4.2.2		High
4.2.4		High
4.2.5		High
4.1.2		Medium
4.2.1		Medium
4.2.3		Medium
4.2.6		Medium
4.1.5		Low

5 BCM Strategy

5.1 Strategy Requirements

		Risk	Workpaper References
5.1.1 Strategic options have been defined for the auditee organization.		High	
Security Concern/Implication If no options exist that reflect different levels of residual risk, the ultimate management decision regarding risk and risk mitigation may not be clear. As a result, investments and actions may not be in line with business decisions, leading to an imbalanced BCM budget and areas not covered by appropriate funding. **Recommendation** Scenarios, concepts and the overall level of risk tolerated by management should be consolidated into strategic options that clearly outline BCM strategies and residual levels of risk.	**Manual Review Procedure** Review the documentation and check if clear options have been defined in a strategic sense. There should be no more than three or four options expressing the different strategic levels of risk mitigation vs. risk taking. **Automated Tool Reference** N/A		BCM strategy overview document supporting material for strategic options

Detailed audit activities to be performed:

- Check if the results of the business impact assessment (BIA) have been consolidated into a number of options outlining the elements of risk mitigation, risk taking and corresponding technical / managerial implications.
- Verify if a decision support matrix has been prepared showing all options in a clear and concise manner. Strategic options should be clearly distinguished.
- Verify that all strategic options have been evaluated in terms of expected cost and benefits. Options should be validated and analysed with a view to providing clear guidance to management as regards required investments, benefits and risk reduction levels.

		Risk	Workpaper References
5.1.2 Technology and non-technology issues have been covered by the strategic options.		Medium	
Security Concern/Implication If the strategic options do not address technical and non-technical issues in a comprehensive manner, the implications of selecting an option may not be transparent. Management decisions may therefore be taken under conditions of uncertainty, with an insufficient understanding of risks and consequences involved. **Recommendation** The strategic options should outline non-technical and technical issues in a comprehensive manner, taking into account financial and other consequences.	**Manual Review Procedure** Verify documentation and obtain an overview of strategic options. Ensure that both dimensions (technical and non-technical) have been considered. **Automated Tool Reference** N/A		as above

Detailed audit activities to be performed:

- Review the documents used to prepare the strategic options. Check that the options are not restricted to IT or other technical fields within the auditee organization.
- Verify that technology has been covered at an adequate level. Consult external experts where appropriate. Use technical audit manuals if necessary to form a view of the plausibility of the strategic options.

Some technological environments may require knowledge not available to the audit team, particularly in complex manufacturing or other industrial organizations. Likewise, a heavily IT-dependent environment may not lend itself to a direct assessment. If in doubt about the ability to adequately cover the technological situation, consult experts in the relevant fields.

		Risk	Workpaper References
5.1.3 External (third party) products and services have been included in the strategic options.		High	
Security Concern/Implication If strategic options do not address third party products and services, there may be no continuity planning for these. This means that disruptions and outages might occur whose consequences are beyond the control of the company. **Recommendation** All third party products and services should be included in the strategic options to ensure continuity in the event of a crisis or disaster.	**Manual Review Procedure** Verify that all third party products and services have been reviewed in a strategic sense. Recursively audit the suppliers and business partners to ensure that they themselves have set up strategic options for business continuity. **Automated Tool Reference** N/A		

Detailed audit activities to be performed:

- Obtain the list of third party products and services used by the company or entity reviewed. Check for completeness. Make sure that the levels of criticality for third parties have been clearly established.
- Referring to section 3 of this program, validate the business impact assessment for third parties. If any outsourced (critical) products or services exist, they should form a major part of the strategic options. Non-critical external resources should still be mentioned in strategic options addressing prolonged disruptions or outages.

		Risk	Workpaper References
5.2 BIA Alignment			
5.2.1 All strategic options have been aligned with the results of the business impact assessment (BIA).		Medium	
Security Concern/Implication If strategic options are not formed in accordance with BIA results, they may not address business functions in terms of their criticality. Investments and managerial attention may therefore be diverted to less critical processes, leading to dysfunctional plans or higher damage / impact in the event of a disruption. **Recommendation** All options considered should be clearly linked to the results of the business impact assessment. Criticality ratings, possible impacts and the resulting level of required action identified as the result of the BIA should be the main element in formulating strategic options.	**Manual Review Procedure** Compare BIA results with the tableau of strategic options. Ensure that impact-based decisions have been included in the formulation of options. **Automated Tool Reference** N/A		

Detailed audit activities to be performed:

- Refer to section 3 (BIA) of this audit program. Verify BIA results and familiarize with the critical processes and areas of maximum impact.
- Compare the BIA results with the set of options identified by the company or entity under review. Check whether all options have a clear and unambiguous link to the critical resources to be protected. Highlight any options or subsets of options that do not address resources and processes in their order of criticality. Follow the way in which an option proposes to allocate scarce resources in the event of a crisis or disaster: are critical resources (in the sense of the BIA) protected first? What is the order of recovery and restoration envisaged by the option?

<table>
<tr><td colspan="3"></td><td>Risk</td><td>Workpaper References</td></tr>
<tr><td></td><td colspan="2">5.2.2 Cost-benefit calculations have been performed to validate the strategic options in terms of viability and residual risk.</td><td>Medium</td><td></td></tr>
<tr><td></td><td><u>Security Concern/Implication</u>
If the cost for implementing an option exceeds the maximum potential damage, this invalidates the option as such. If residual risk is too high, the option may not provide a reasonable level of assurance in the sense of prudent management.

<u>Recommendation</u>
All options should be quantified in terms of cost and effort. These should be weighed against the benefits derived from implementing the options. The final decision for a strategic option should be based on these cost-benefit calculations.</td><td><u>Manual Review Procedure</u>
Review cost calculations for implementation of the various options identified. Review potential damage (without any preparation or business continuity). Review the residual risk of damage (after the option has been implemented). Assess the plausibility of the options outlined.

<u>Automated Tool Reference</u>
General office software such as Microsoft Excel or Corel Quattro Pro</td><td></td><td>cost calculation sheets</td></tr>
</table>

Detailed audit activities to be performed:

- Compile a detailed documentation of cost calculations for the strategic options identified. This should include assets, expenditure and third party costs for implementing, maintaining and improving the detailed steps within each option, over a reasonable period of time.
- From the BIA results, calculate the maximum potential damage that the entity is likely to face when experiencing a disruption or disaster without any preventive measures or preparation, i.e., without any business continuity management.
- Taking into account the reduction in potential damage that is achieved by implementing the strategic options (consider each option separately), compare the investment for the option with the reduction in risk. Highlight any instances where the proposed investment is equal to or higher than the maximum potential damage.
- Identify any options that lead to a high level of residual risk, i.e. "minimalist" solutions. If the reduction in potential damage is negligible, the option may not go far enough in terms of protecting the organization.
- Ensure that similar calculations and strategic decision processes have taken place within the auditee organization. Perform independent cost-benefit analyses on a sample basis where appropriate. Check the methods used for deriving cost and benefit values.

5.3 Outsourcing / Insourcing Issues		Risk	Workpaper References
5.3.1 Options have been evaluated in terms of insourced vs. outsourced activities.		High	
Security Concern/Implication If insourcing vs. outsourcing has not been evaluated for each strategic options, hidden risks may exist that render business continuity management dysfunctional. Significant outsourcing risks may not be recognized. **Recommendation** Any insourcing / outsourcing decisions that are envisaged as part of strategic options should be evaluated in detail, particularly with regard to hidden outsourcing risks.	**Manual Review Procedure** Review the documentation for strategic options. Where outsourcing or insourcing (if outsourced before) decisions are envisaged, check that relevant risk evaluations have been performed. **Automated Tool Reference** N/A		

Detailed audit activities to be performed:

- Obtain all proposed outsourcing agreements (hot site, IT applications, back-end business functions, infrastructure functions, security etc.) and corresponding service level agreements. Likewise, examine in detail all processes and functions that are currently outsourced and planned for insourcing as part of a strategic option.
- For outsourced activities, verify that the outsourcer´s BCM capability is at least equal to that of the auditee organization. Check whether ServiceLevel Agreements (SLAs) contain clauses indicating a "buy-in" to the organization's business continuity management practices.
- Verify that the outsourcing partner has understood the BCM concept. Conduct interviews where necessary to ensure that the outsourcing partner is aware of the contractual liabilities and the principal's requirements for the product or service in question. Finally assess whether the outsourcing partner is likely to be able to carry out the business process or processes under adverse circumstances.
- For insourced activities, verify that the skills and knowledge exist to perform the process or processes in-house. Assess the mechanism for insourcing (in the event of a crisis or disaster !) and find out whether subsequent reversal of such arrangements is envisaged. Check SLAs with the current outsourcing partner to ascertain contractual arrangements, remedies and any descriptions showing that both partners clearly understand the significance and magnitude of these events.
- Both outsourcing and insourcing plans that form part of strategic options should be seen in terms of cost and benefit (see this section of audit program). Highlight any discrepancies and any plans that are not reasonably justifiable on a cost-benefit basis.

		Risk	Workpaper References
5.3.2 A service level strategy has been envisaged for outsourced products and services that form part of the predetermined strategic options.		High	
Security Concern/Implication If outsourced products and services are not governed by a service level framework, they may not be available in the event of a crisis or disaster. **Recommendation** All outsourced activities should be covered by service level agreements in line with overall business continuity management requirements.	**Manual Review Procedure** Review outsourcing agreements and service level documentation where available. Verify that service levels are in line with the requirements set down in business continuity plans. **Automated Tool Reference** N/A		outsourcing and service level agreements

Detailed audit activities to be performed:

- Review all outsourcing agreements: check if service levels have been defined for normal operations and for alternative operations.
- Ensure that service levels have been set in accordance with BCM requirements. This should include maximum tolerable downtimes, reaction and response times.
- Verify that all outsourcing partners are "in the loop" in terms of emergency response and immediate actions following a disruptive event: call trees, notification, crisis teams, etc. The outsourcing partner should form an integral part of the organization's crisis management capability.
- Examine in detail whether service levels and outsourcing agreements provide appropriate remedies in contractual, insurance and legal terms. Consult external experts where necessary.

5.4 Enterprise-wide Strategy		Risk	Workpaper References
5.4.1 Strategic options are designed to cover the whole organization.		High	
Security Concern/Implication If strategic options do not cover the whole organization, "blank" areas may exist where no business continuity has been envisaged. This may lead to increased exposure and a higher level of potential damage. **Recommendation** Strategic options should cover all parts of the organization and its business processes, as well as subsidiaries, affiliated companies and relevant third parties.	**Manual Review Procedure** Review an overall map of the auditee organization and verify that all parts have been included in the strategic options. **Automated Tool Reference** N/A		

Detailed audit activities to be performed:

- From organization charts and other standard documentation, derive a mapping of all entities belonging to the auditee organization. These may be prioritized, geographically ordered or otherwise categorized.
- Verify that all parts of the map are covered by strategic options. Typically, BCM is extended to small parts (central locations, big subsidiaries, headquarters, etc.) only, depending on the priorities assigned by group management. Examine these priorities. Highlight any weaknesses in the concept. Ensure that, if a phased approach to BCM has been decided upon, senior management has set down a clear roadmap towards including the "second wave" and subsequent phases, to include all entities, affiliates etc.
- If strategic options have been defined for a part of the organization, check their plausibility with a view to extending the concepts to other parts. Highlight any discrepancies, for instance where strategic options for headquarters cannot be transferred to outlying subsidiaries. If this is the case, different strategies should exist for each type of corporate unit or subsidiary – identify any weaknesses or entities not covered by appropriate BCM strategies.

		Risk	Workpaper References
5.4.2 The strategic options subdivide plans into corporate, business unit and department levels according to requirements set by the BIA.		Medium	
Security Concern/Implication If the strategic options do not include the normal organizational structure, they may not be suitable for all units and departments. This may lead to increased exposure and dysfunctional plans in parts of the overall organization. **Recommendation** Strategic options should take into account the structure of the organization, covering business units, departmental structures and other discrete units.	**Manual Review Procedure** Review organization charts and other structural maps. Ensure that all parts have been recognized in a structural sense, by extending the strategic options to that type of unit / department. **Automated Tool Reference** N/A		

Detailed audit activities to be performed:

- Review the description of options. Check if all types of organizational unit (subsidiaries, divisions, groups, departments, sub-departments etc.) have been addressed, either directly or indirectly. Typically, some references will have been made to "departmental plans" or other sub-documents linking into the strategy. If this is not the case, highlight the inconsistencies and open issues.
- If a BIA (see above) has been performed, the granularity used there should be mirrored by the strategic options. For instance, if criticality ratings for business processes have been prepared in cooperation with department heads, the resulting strategic options should acknowledge department heads as key players in the detailed strategy for BCM.
- Units of responsibility for BCM (subsidiary, department, group etc.) should match units of strategy. Those responsible for owning and maintaining business continuity should be the main providers of input for the strategy. Highlight situations where (as an example) the BIA has been prepared by interviewing department heads, but strategy and budgeting are done at the divisional level, as the two levels might pursue different objectives.

		Risk	Workpaper References
5.4.3 The BCM strategy assigns responsibilities and tasks at the appropriate organizational levels.		High	
Security Concern/Implication If responsibilities and tasks are not assigned at appropriate organizational levels, there may not be enough resources or expertise to ensure proper performance and maintenance. Plans may be dysfunctional or out of date. **Recommendation** All responsibilities and resulting tasks should be assigned to the appropriate organizational levels and individuals.	**Manual Review Procedure** Review tasks lists and defined areas of BCM responsibility. Map these against organizational charts. **Automated Tool Reference** N/A		task lists BCM team descriptions organizational charts

Detailed audit activities to be performed:

- Review overall objectives of the BCM strategy, work packages and individual task lists.
- Review the organizational structure for BCM – teams, department representatives, BCM experts, country representatives etc.
- Compare the two structures. Highlight any areas where the granularity of BCM does not match the level at which it is assigned. Typically, such discrepancies might be in evidence where a part-time IT manager is assigned country responsibility for BCM, or where a technical department is assigned the responsibility of BCM for core business processes.
- Examine in detail how senior management (group, division, subsidiary) is involved in BCM-related activities. Highlight any elements of the BCM strategy without senior management participation.

Overview Chapter 5 Audit Items

Number	Preliminary Rating Assigned by Auditor (from 1 to 5, 1 = absent or inadequate, 5 = comprehensive and adequate), and auditor's comment	Risk
5.1.1		High
5.1.3		High
5.3.1		High
5.3.2		High
5.4.1		High
5.4.3		High

5.1.2		Medium
5.2.1		Medium
5.2.2		Medium
5.4.2		Medium

6 Detailed Business Continuity Planning

6.1 Plan Development Requirements

6.1.1 The planning organization, teams and job descriptions are defined and documented.		Risk	Workpaper References
		Medium	
Security Concern/Implication If the business continuity planning organization, the teams and their job descriptions are not defined and documented, confusion and misunderstandings may arise. As a result, plans may be incomplete, conflicting or dysfunctional. **Recommendation** The planning organization and planning teams should be clearly defined, as should be job descriptions and planning tasks.	**Manual Review Procedure** Review organization charts of the planning process, the teams responsible and job descriptions of individual team members. Highlight any weaknesses. **Automated Tool Reference**		organization charts job descriptions team descriptions

Detailed audit activities to be performed:

- Obtain an overview of the organization structure for business continuity planning. This may be different from the structure envisaged for a real disaster or emergency. Review the organization chart for completeness. Check that all planning functions have been covered.
- Review team descriptions across organizational units. Assess adequacy of team size and functionality. Highlight any weaknesses in local planning teams.
- Obtain job descriptions or descriptions of roles and responsibilities. Compare these for several of the teams reviewed. Assess individual tasks and accountability.

		Risk	Workpaper References
6.1.2 Planning aids, forms and an appropriate information database have been included in the detailed planning concept.		Low	
Security Concern/Implication If no planning aids, information databases and standardized forms are provided for planning, the resulting plans may be confusing and incorrect in terms of background information **Recommendation** Appropriate planning aids, forms and a suitable information database should be provided for planning.	**Manual Review Procedure** Review the list of planning aids. Sample some forms and other tools and review for adequacy. **Automated Tool Reference** Optional information database for planning.		

Detailed audit activities to be performed:

- Obtain a list of planning aids, including the planning guidelines, standardized forms and a description of the information database used.
- Review the guidelines for completeness and adequacy. Guidelines should cover all stages of the planning process and explain what the objectives of planning are.
- Review sample forms for planning. Ensure that the standardization process for these forms has been adequate, i.e., that no "fill-in-the-forms" approach has been taken. Highlight any weaknesses in forms or in the way they are utilized. Obtain additional forms for review if in doubt.
- If an information database is used, review its design and deployment throughout the planning organization. Test whether planning teams are aware of the database and reasonably skilled in using it. Check that the database is up to date. Refer to *chapter 12* of this section if the database is part of a larger BCM tool.

Note: it is possible, particularly in smaller organizations, that planning takes place without standardized planning aids. If this is the case, mitigating controls (such as frequent meetings of the planning groups) should be in place. In larger organizations, this practice might become cumbersome and difficult to implement. Where a process of standardization has been introduced, it should be scrutinized for signs of being mechanistic, i.e., the process should not allow planners to cut corners and just fill in the forms. If a database is utilized and relied upon, it should be up to date and maintained properly. In these cases, frequent access to and frequent updating of the database should be verified by means of suitable tests.

			Risk	Workpaper References
	6.1.3 A comprehensive package for planning has been distributed to all business continuity coordinators and other individuals involved in planning.		Low	
	Security Concern/Implication If no standardized approach for planning has been communicated to the coordinators and other planners, the resulting plans may not follow the agreed standard. **Recommendation** A comprehensive planning package should be distributed to all business continuity coordinators and other planners.	**Manual Review Procedure** Obtain a sample of the BC package. Review the documentation and highlight any missing items or weaknesses in the contents. **Automated Tool Reference** N/A		BC package

Detailed audit activities to be performed:

- Obtain a copy of the standardized business continuity planning package. Verify that the same package has been distributed to all BC coordinators and other planning team members. If different sets of packages have been used, enquire about the reasons. In this case, you should verify all packages that have been circulated.
- Analyze the contents of the package in terms of structure, ease of use, and completeness. Where different packages are in evidence, analyze their adequacy for the audiences they are intended for.
- Highlight missing items or structural deficiencies. The package should enable any reasonably knowledgeable individual to commence planning in the standardized format agreed beforehand.

<table>
<tr><td colspan="3"></td><td colspan="2"></td></tr>
<tr><td colspan="3">6.2 Recovery Management and Control Requirements</td><td>Risk</td><td>Workpaper References</td></tr>
<tr><td></td><td colspan="2">6.2.1 Stages of escalation and trigger events (interruption, crisis, disaster) are clearly defined and used as the basis for detailed business continuity plans.</td><td>High</td><td></td></tr>
<tr><td></td><td><u>Security Concern/Implication</u>
If no escalation triggers have been defined, there may be confusion as to when and how to escalate an incident to the stage of "crisis" or "disaster." There is a risk of increased damage to the organization. Recovery may be impeded if the critical condition is not recognized, or recognized too late, as a result of deficient escalation triggers.

<u>Recommendation</u>
Trigger events should be defined for escalation of an event to the next stages of plan activation and initiating business continuity measures.</td><td><u>Manual Review Procedure</u>
Obtain a list of trigger events and other escalation procedures. Review in detail and report weaknesses.

<u>Automated Tool Reference</u>
N/A</td><td></td><td>List of trigger (Escalation) events</td></tr>
</table>

Detailed audit activities to be performed:

- Obtain a list of escalation procedures and "trigger events" (these may be given a different name in the organization being reviewed).
- Check that all events are validated against the scenarios and events in the list. There should be clear indicators explaining to the planner or other person in charge when and how to escalate an event or incident to the next stage envisaged in the business continuity plan.
- Events triggering a partial or full activation of emergency measures and business continuity plans should be defined in a concise, unambiguous and clear manner. You should be able to understand them immediately, and there should be no room for questioning them.

			Risk	Workpaper References
	6.2.2 Escalation criteria are set and have been communicated to all planners.		High	
	Security Concern/Implication If no escalation criteria have been defined, set down and communicated, there may be confusion as to how to escalate, to whom and when. There is a risk that the business continuity plan is not activated or activated too late to contain the event. **Recommendation** Escalation criteria should be clearly defined, documented and communicated to all parties concerned.	**Manual Review Procedure** Obtain a copy of the escalation criteria and related procedures. Check for adequacy and realism. **Automated Tool Reference** N/A		Escalation criteria (list) Escalation procedures

Detailed audit activities to be performed:

- (*also refer to 6.2.1 above*) Obtain and review a list of escalation criteria. This should include procedures describing what to do, when to do it and to whom to report. The criteria may be quite closely linked to the events or conditions triggering them.
- Make sure that the escalation process is reasonable and realistic. Typical pitfalls as below should be avoided at all cost:
  - Non-escalation for fear of having to face consequences if it was a false alert
  - Partial escalation due to incomplete information about the event
  - overlooking a trigger event that is in the list
  - Premature escalation due to misunderstandings of criteria or trigger events
  - Incorrect decisions of superiors due to partial or irrelevant information being escalated
  - Inaction in cases where an event is "not on the list"
  - Overly rigid escalation criteria
  - Overly loose escalation criteria

Note: this item is not just related to detailed BCM, but also to chapter 4 of this section. The auditor should dwell on this subject if necessary, to ensure absolute clarity with regard to the overall escalation process. If this audit item shows significant weaknesses, it may invalidate all subsequent phases of managing the crisis, recovering and restoring normal operations. Delayed or incomplete escalation may cause considerable damage which would have been avoidable. As an auditor, it is recommended that you take a "better safe than sorry" approach towards reviewing escalation mechanisms, trigger events and procedures.

		Risk	Workpaper References
6.2.3 Recovery teams have been defined and described at an adequate level of detail. Roles and responsibilities are clearly assigned to individuals.		**High**	
Security Concern/Implication If no recovery teams have been defined, or if the description is incomplete, recovery after a critical event may be delayed or deficient. **Recommendation** Recovery teams should be defined and described at an adequate level of detail, including roles and responsibilities that are assigned to individuals.	**Manual Review Procedure** Review recovery team descriptions and related material for completeness and adequacy. **Automated Tool Reference** N/A		Recovery team structure (charts) Team descriptions Roles and responsibilitie s (documents)

Detailed audit activities to be performed:

- Review the standard format of definition and description for recovery teams. Include organization charts, documents and any automated mechanisms such as personnel databases (refer to *chapter 12* of this section where necessary).
- Verify that team structure, individual roles and responsibilities are well defined and easily understood by team members.
- Make sure that the recovery team structure is uniform throughout the organization. There should be no divergence between teams for different organizational units, unless there are clear reasons for this (such as different business operations). The corporate, higher-level teams should be uniform so that any team member from one team can find their counterpart in another recovery team.

		Risk	Workpaper References
6.2.4 The planning concept clearly distinguishes between emergency response, recovery and restoration phases.		Medium	
Security Concern/Implication If phases of business continuity have not been defined, the activation of teams and resources for each phase may be delayed. As a result, increased damage or flawed recovery efforts may occur. **Recommendation** The planning concept should clearly outline different phases of business continuity management such as emergency response, recovery and restoration of normal operations.	**Manual Review Procedure** Review the overall business continuity concept, looking for a phased approach. Sample individual plans and compare the phases. **Automated Tool Reference** N/A		

Detailed audit activities to be performed:

- Find out whether the overall BC concept contains phases of management for incidents and critical events. The phases may follow a typical "disaster timeline" approach, ranging from normal operations (before the event) to restoration of normal operations.
- Sample detailed plans and check that these phases have been reiterated in the plans. All planners, team members and BC personnel should be aware of the phases.
- The phased approach should be communicated and well understood by all involved. Highlight any weaknesses with regard to the lack of a phased approach and / or lack of communication.

Note: emergency response, recovery to an interim state of operations, and restoration to normal operations are an intrinsic concept to BCM. These phases may be given different names in the organization being reviewed, but their underlying significance should be clear. You should make sure that the phases are designed in a way that allows timely activation of different teams and resources, depending on what is needed at that moment in time. The "disaster timeline" may be seen as a form of red line connecting the activities of different teams, external suppliers etc.

6.3 Format and Structure of Plan Components		Risk	Workpaper References
6.3.1 All business continuity plans follow an integrated and agreed structure.		High	
Security Concern/Implication If there is variance in the structure of detailed BC plans, they may be difficult to compare, understand, maintain and use in the event of a disaster. This will increase the risk of delayed or flawed recovery. **Recommendation** All plans should follow an integrated structure that has been agreed beforehand and communicated to planners.	**Manual Review Procedure** Review plan structure documents ("architecture"). **Automated Tool Reference** N/A		

Detailed audit activities to be performed:

- Where available, obtain information on how plans are structured (either top-down or in any other format). Review the "architecture" of plans, for instance the document structure for technology plans and its integration with managerial plans.
- Examine in detail any cross-references between plans and highlight weaknesses. Cross-references are a good indicator of an integrated or not sufficiently integrated plan document structure.
- Obtain and check documents that show prior agreement between BC management, planners and line departments. All parties involved should have agreed on the integrated planning document "architecture."

			Risk	Workpaper References
	6.3.2 A comprehensive control and reporting system has been implemented for detailed business continuity plans.		Medium	
	Security Concern/Implication If no control and reporting system exists, information exchange between BC coordinators and planners may be delayed. Rework of plans may result. There may be misunderstandings with regard to detailed plans. **Recommendation** A comprehensive control and reporting system should be in place for detailed business continuity planning.	**Manual Review Procedure** Obtain and review control and reporting procedures for the planning phase. **Automated Tool Reference** N/A		control and reporting procedures

Detailed audit activities to be performed:

- Conduct a detailed review of the control and reporting system for BC planning. It should cover a) creating plans; b) modifying plans; c) reporting on planning progress; and, d) approval and finalization of plans.
- If the BC coordinators do not receive adequate, timely information at all times, this should be noted as a weakness of the control and reporting system. Likewise, BC coordinators should be able to immediately report on overall planning progress on the basis of the reporting system.

		Risk	Workpaper References
6.3.3 Appropriate templates and sample files have been set up for reference purposes.		Medium	
Security Concern/Implication If no templates and samples are available, this may lead to misunderstandings and incorrect planning. **Recommendation** Appropriate templates and sample files should be available for reference purposes.	**Manual Review Procedure** Review templates and samples. Highlight weaknesses in contents as well as distribution or accessibility. **Automated Tool Reference** N/A		Templates Sample files

Detailed audit activities to be performed:

- Obtain copies of templates and sample files. These may be part of the BC planning package (see above). Review contents.
- Examine the distribution list and check accessibility of the templates and sample files, e.g. ,on the corporate intranet or through e-mail request. Difficulty in obtaining the files in this way should be reported as a weakness.
- Verify that samples and templates are reasonably clear. They should cover a representative selection of departmental, business function, and technology examples for planning.

Note: as mentioned above (BC planning package), samples and templates may not always be a good idea, particularly where planners adhere to them too closely. You should carefully examine the materials provided and determine whether they encourage a "fill-in-the-forms" approach on part of the planner. If this is the case, take representative samples of departmental plans and look for "copy-and-paste" type parts in them. If these are in evidence, the sample files / template approach is weak and requires closer attention.

6.4 Operational Planning		Risk	Workpaper References
6.4.1 All operational areas and business units are covered by detailed plans.		High	
Security Concern/Implication If some areas and business units are not covered by detailed plans, they may not be able to recover from a crisis or disaster. There is a risk of considerable damage and subsequent inability to conduct business. **Recommendation** All operational areas and business units must be covered by detailed plans.	**Manual Review Procedure** Review the reporting on plan completion and approval. Highlight any missing items. **Automated Tool Reference** N/A		

Detailed audit activities to be performed:

- Obtain appropriate reporting sheets and approval minutes for all business units. Match these against an organization chart.
- Identify and specifically report any missing plans. Examine in detail why the plans are missing. If they are delayed, report this as a weakness. If they have not been prepared at all, report this as a major weakness of the BC concept as a whole.
- Depending on the department, area or business unit that does not have a plan, inform senior management of the finding. If critical operational departments are without plans, this should prompt immediate management action.

		Risk	Workpaper References
6.4.2 The plan Table of Contents is comprehensive and easy to use.		Low	BCI / DRII Professional Practices
Security Concern/Implication Without a Table of Contents that is comprehensive and easy to use, personnel may not be able to locate specific information and the plan may be difficult to follow. **Recommendation** We recommend that the business continuity plan contain a Table of Contents which is representative of the text and is easy to use.	**Manual Review Procedure** Review the Table of Contents in the business continuity plan to ensure that it is accurate and easy to use. Assess completeness, against known BCM models. **Automated Tool Reference** N/A		

Detailed audit activities to be performed:

Match the TOC against known BCM / BCM models and check completeness. Among others, the following models are available as examples (typical generic models):

- Business Continuity Institute (BCI) knowledge areas
- Disaster Recovery Institute International (DRII) method
- Strohl Systems (LDRPS) Business Continuity Planning Guide

			Risk	Workpaper References
	6.4.3 Emergency evacuation procedures, fire regulations, and other building safety procedures are included in the plan and tested.		Medium	High priority is assigned to the detailed shape and formal plan as mandated by standards and laws (see section 2)
	Security Concern/Implication If there are no evacuation and safety procedures included in the plan, there is a potential for confusion during a disaster which could lead to personnel being injured or unable to escape. **Recommendation** We recommend including detailed emergency evacuation procedures, fire regulations, and other building procedures in the plan. These safety procedures are to be tested and updated as required.	**Manual Review Procedure** Query the appropriate party and/or parties to determine the necessary documentation to ensure safe evacuation and recovery for all personnel. Document the procedures in the plan if they are not present. If present, review for accuracy. **Automated Tool Reference** N/A		

Detailed audit activities to be performed:

- Check evacuation procedures and manually go through the motions (walk along escape routes, try doors etc.).
- Inspect mustering stations or other areas that people are assigned to. Do not test emergency doors where sealed or linked to an electronic alarm system.
- Check headcount procedure (after evacuation)
- Obtain information about earlier exercises (e.g., fire evacuation drill) and results. Verify if fire drills are performed on a regular basis.
- Check if fire extinguisher training has been offered / ordered.

		Risk	Workpaper References
6.4.4 A plan distribution list is included in the plan.		Low	
Security Concern/Implication Not including a plan distribution list in the plan could result in the inability to properly distribute updates to all personnel involved in the recovery efforts. **Recommendation** We recommend including a plan distribution list in the plan.	**Manual Review Procedure** Review the business continuity plan to determine if a distribution list is included in the plan. **Automated Tool Reference** N/A		

Detailed audit activities to be performed:

- refer to *chapter 6.6* (plan distribution and control) for audit steps

Note: if other forms of distribution and control are in place, it is not essential that the plan itself contain a distribution list. Hence, the risk is classified as "low." If the plan distribution and control mechanism as a whole is weak, this finding will be more significant.

		Risk	Workpaper References
6.4.5 All computing recovery requirements (i.e., number of servers, space on servers, memory on servers) have been determined and are documented in the plan, and arrangements to obtain this equipment are detailed in the business continuity plan.		Low	
Security Concern/Implication If arrangements are not made to obtain computing equipment when a disruption occurs, it is possible that recovery could be slowed or made impossible. If computing requirements are not determined and documented, it is possible that incorrect computing equipment will be supplied, thus hindering an efficient recovery **Recommendation** We recommend that all computing requirements are determined and documented in the business continuity plan, and arrangements have been made to obtain this equipment when it is needed.	**Manual Review Procedure** Review the business continuity plan to determine that the computing recovery requirements have been determined and documented in the plan. Ensure that all the pertinent information is included, such as number of servers & PCs, disk space needed, RAM needed, special computing requirements, etc. Review the plan to also ensure that arrangements have been made to obtain the proper equipment if a disaster were to occur. **Automated Tool Reference** N/A		

Detailed audit activities to be performed:

- Refer to materials and methods on IT disaster recovery audit. Distinguish between ad hoc, workplace computing equipment such as notebook computers, and larger IT infrastructures.
- This BCM audit is NOT concerned with IT disaster recovery steps taken by the organization. Limit your audit activity on this point to the bare minimum needed by the crisis management team, the EOC, etc. Do not attempt to perform an in-depth examination of IT recovery.

Note: it is suggested that you review the IT disaster recovery concept (if existing) very briefly and at a high level. If this concept appears appropriate, no further audit steps should be performed. If there is reason to believe that IT-related continuity steps may be deficient, highlight this as a high level finding ("IT disaster recovery does not appear to be comprehensive") and recommend that a separate audit be carried out to clarify the situation. Even if you feel well equipped and knowledgeable enough to tackle the IT side of things (perhaps as a CISA – Certified Information Systems Auditor), keep in mind that this is a BCM budget which is limited.

			Risk	Workpaper References
	6.4.6 All necessary support equipment (copy machines, fax machines, etc.) needed for recovery has been determined and documented in the plan, and arrangements to obtain the equipment are detailed in the business continuity plan.		Medium	
	Security Concern/Implication If arrangements are not made to obtain the necessary support equipment when a disruption occurs, it is possible that recovery could be slowed. If support equipment requirements are not determined and documented, it is possible that incorrect equipment will be supplied, thus hindering an efficient recovery. **Recommendation** We recommend that all necessary support equipment is determined and documented in the business continuity plan, and arrangements have been made to obtain this equipment when it is needed.	**Manual Review Procedure** Review the business continuity plan to determine that the support equipment recovery requirements have been determined and documented in the plan. Ensure that all the pertinent information is included, such as any particular brand needed, color copiers/printers needed, quantity needed, etc. Review the plan to also ensure that arrangements have been made to obtain the proper equipment if a disaster were to occur. **Automated Tool Reference** N/A		equipment lists service level agreements with suppliers

Detailed audit activities to be performed:

- Obtain lists of necessary support equipment. Compare against the amount of equipment used during normal operations. Check if the number of appliances and items required for alternative operations is adequate.
- Check SLA's with equipment suppliers. Ensure that the delivery and response time are adequate.

			Risk	Workpaper References
	6.4.7 All special forms and supplies needed for recovery are identified, and arrangements to obtain them are documented in the business continuity plan.		Medium	
	Security Concern/Implication If arrangements are not made to obtain the special forms and supplies when a disruption occurs, it is possible that recovery could be slowed. If special forms and supplies needed for recovery are not identified and documented, it is possible that recovery will not be timely. **Recommendation** We recommend that all special forms and supplies needed for recovery are identified and documented in the business continuity plan, and arrangements have been made to obtain these forms and supplies when they are needed.	**Manual Review Procedure** Review the business continuity plan to determine that the special forms and supplies have been identified and documented in the plan. Review the plan to also ensure that arrangements have been made to obtain these forms and supplies if a disaster were to occur. **Automated Tool Reference** N/A		

Detailed audit activities to be performed:

- Review off-site storage of supplies and essential forms such as letterhead papers. Ensure that an adequate stock is held, either by way of "back-to-back" arrangement between subsidiaries, or at a third party location.
- If no stock levels have been envisaged, check that the supplies and forms can be reproduced from electronic templates, or ordered from suppliers.

6.4.8 All voice and data communications equipment needed (i.e., switches, multiplexors) for recovery in order to support the critical business processes is identified, and backup provisions for this equipment are detailed in the business continuity plan.		Risk	Workpaper References
		Low	
Security Concern/Implication If arrangements are not made to obtain communications equipment when a disruption occurs, it is possible that recovery could be slowed or made impossible. If communications equipment requirements for recovery are not determined and documented, it is possible that incorrect equipment will be supplied, thus hindering an efficient recovery. **Recommendation** We recommend that all communications equipment needed for recovery are determined and documented in the business continuity plan, and that arrangements are made to obtain this equipment when it is needed.	**Manual Review Procedure** Review the business continuity plan to determine that the communications requirements for recovery have been determined and documented in the plan. Ensure that all the pertinent information is included. Review the plan to also ensure that arrangements have been made to obtain the proper equipment if a disaster were to occur. **Automated Tool Reference** N/A		

Detailed audit activities to be performed:

- standard terrestrial voice comms (analog / ISDN / other stationary telecomms)
- LAN / WAN cabling and interfaces, incl. external service providers
- wireless / GSM comms / GPRS / Other
- capacity planning, alternative operations
- load balancing and redirecting arrangements for alternative locations
- preference and precedence arrangements with providers and / or state-owned networks
- equipment lists, integrity of concept, feasibility, fallback positions etc. for alternative operations should be reviewed in detail.
- compare against increased load conditions on networks whilst in alternative ops (heavy usage)

Note: as with IT equipment, this exercise should be strictly limited to the immediate needs of the organization's vital functions during recovery, i.e., the EOC, senior managers and BC personnel. It is not the objective of this audit program to conduct a full-scale review of telecommunications recovery. Bear in mind that this is another very complex technology field that may require external assistance and a separate and distinct budget. As mentioned above (see IT equipment recovery in part 6.4 of this section) a high-level review of the telecommunications situation should be carried out, reporting any weaknesses as a summary, and summarily recommending an in-depth audit of telecommunications as appropriate.

		Risk	Workpaper References
6.4.9 Back-up tapes are run nightly, stored off-site, and labelled for easy retrieval.		Low	
Security Concern/Implication If back-ups are not properly performed and easily retrievable, the systems may be difficult to restore in a timely manner. **Recommendation** We recommend that all systems are backed-up on a regular basis and the tapes are stored off-site and labelled for easy retrieval.	**Manual Review Procedure** Query the system administrator as to whether back-ups are being performed on a regular basis. If they are, discuss where the tapes are being stored and the labeling procedures. **Automated Tool Reference** N/A		

Detailed audit activities to be performed:

- Check if a backup and restore concept exists.
- Verify that the necessary backups will be available as part of the appropriate phases
- Check the service level agreement with an off-site storage provider (if applicable)

Note: Assume that the backup concept has been finalized, tested and kept up to date as part of the IT disaster recovery plan. This is not a part of this audit. Where weaknesses are found, recommend a separate review as part of a full IT disaster recovery audit.

	Risk	Mitigating Controls & Workpaper References
6.4.10 A list of all personnel to be involved in the recovery effort, with their office and home phone numbers, is included in the business continuity plan.	**Medium**	
Security Concern/Implication Failure to document participants and phone numbers in the business continuity plan could result in incorrect or missing information, which could slow the recovery effort. **Recommendation** We recommend maintaining a list of all personnel to be involved in the recovery effort, which includes their names, office and home phone numbers.		
Manual Review Procedure Review the business continuity plan to ensure that a current list of personnel involved in the recovery effort, along with their phone numbers, exists. Cross reference this list with the team members listed for each business process to ensure that all personnel are included. **Automated Tool Reference** N/A		

Detailed audit activities to be performed:

- Verify if contact details are accurate and up to date.
- Where personal details are absent, take into account prevailing law on privacy and data protection
- Confirm a process is in place to continually update and validate contact details

		Risk	Workpaper References
6.4.11 All vendor/supplier names and phone numbers are included in the business continuity plan.		**Low**	
Security Concern/Implication If the vendor/supplier names and phone numbers are not included in the business continuity plan, critical business processes may not be recovered in a timely manner due to personnel taking extra time to track down the vendor/supplier information. **Recommendation** We recommend that vendor / supplier names and numbers are included in the business continuity plan in order to minimize the amount of effort and time required at time of disruption.	**Manual Review Procedure** Review the business continuity plan to verify that the vendor/supplier names are included in the plan and are current. **Automated Tool Reference** N/A		contact lists

Detailed audit activities to be performed:

- obtain lists of numbers and directories for third parties who have been identified as part of the business continuity plan
- review the lists and determine whether they are accurate and up to date. Take samples by calling some of the numbers on the lists. Send test faxes to selected numbers.
- Where weaknesses have been identified, drill down to find out what the underlying problems are. Contact the third party organization to verify that their representatives have been adequately informed about the fact that they are included in the business continuity management process.

	Risk	Workpaper References
6.4.12 Team notification procedures (i.e., call trees) are included in the business continuity plan.	Medium	
Security Concern/Implication If team notification procedures are not included in the plan, there is no way to ensure that all key personnel have been notified of the disaster and the next steps they are to take. **Recommendation** We recommend that team notification procedures are included in the business continuity plan. **Manual Review Procedure** Review the business continuity plan to determine if any team notification procedures are included in the plan. **Automated Tool Reference** N/A		

Detailed audit activities to be performed:

- Obtain sample call trees and review for completeness. The call trees should follow a standardized format, containing names, team position and comprehensive contact details.
- Where home numbers or pager details are given, check whether there is a clear chain of calling these numbers.
- Where in doubt, selectively test the contact details given in the call tree listings.
- In countries outside the US, bear in mind that pagers are not very widespread. Cellular phone numbers, call forwarding services and devices alerting the person by transferring calls to short messaging services may be in use.
- In remote regions, and particularly in manufacturing and heavy industrial areas, alternative means of alerting people may be in use. Verify that these satisfy the criteria of immediately alerting the person on the list, and perform appropriate tests.

The sole purpose of a call tree is to reach the person on the list, and to subsequently alert all other persons forming the "tree." The means of doing so may be diverse, depending upon the situation and the company being reviewed. In countries with a limited public infrastructure, alternative means of alerting have been suggested, e.g. the use of audiovisual (tannoy and flashlight) alerting devices. Some situations may even require messengers, ham radio or other devices that may be seen as unusual in highly urbanized environments.

		Risk	Workpaper References
6.4.13 A formal documented agreement exists for the use of an alternate site or other recovery services.		Low	
Security Concern/Implication If formal, documented agreements do not exist for the use of an alternate site or other recovery services, there is a potential for the recovery services/site to be unavailable at the time of a disaster or a recovery test. **Recommendation** We recommend that formal, documented agreements exist for the use of an alternate site or other recovery services.	**Manual Review Procedure** Review the business continuity plan to determine whether any alternate sites or recovery services are being utilised. If they are, query management as to whether formal documented agreements exist for the use of these services. **Automated Tool Reference** N/A		

Detailed audit activities to be performed:

- Review documentation and contracts (detailed)
- Verify precedence / time sharing arrangements: this may be „between the lines".
- Verify that there are clear and unambiguous statements of availability and related timeframes.
- Check against MTD / MTA.
- Verify integration of alternate site contracts with declaration procedures (as reviewed above).

The term "alternative site" has a definite IT bias. While many planners automatically assume that relocation will become necessary in the event of a disaster, this may not be true. As a matter of fact, smaller subsidiaries may successfully operate a "home office" policy or relocate very selectively to other offices, nearby hotels etc. Depending on the size and magnitude of the event for which the plans have been designed, partial relocation may be quite easily effected through using other parts of a large campus.
Some industries may be in a position to effectively help themselves, for instance in the case of construction / logistics business renting containers to form a localized cluster of temporary accommodation. This indicates that an alternate site is one of a number of possible solutions, but definitely not a "must-have."

	Risk	Mitigating Controls & Workpaper References
6.4.14 A recovery solution (hot site, duplicate data center, etc.) has been developed and documented in the business continuity plan.	Low	
Security Concern/Implication If a recovery solution is not developed and a disaster was to occur, it is possible that critical business processes would not be recovered in a timely manner. **Recommendation** We recommend developing an appropriate recovery strategy to utilize in the event a disaster occurs and documenting that strategy in the business continuity plan. **Manual Review Procedure** Review the business continuity plan to ensure that a recovery solution has been developed and documented. If recovery vendors are being used, review all contracts to make sure that they are current. Cross-reference the contracts with the plan to ensure that all necessary recovery equipment is included in the contract. **Automated Tool Reference** N/A		

Contracting a recovery site is usually done when planning specifically for IT or other technological platforms used by the organization. It is therefore part of the corporate IT disaster recovery strategy and not the business continuity strategy proper. Where functions other than IT are to be relocated to an alternate site, refer to item 6.4.15 above.

			Risk	Workpaper References
	6.4.15 For the use of an alternate site or other recovery services, the level of support required from the provider and provisions for obtaining that support are included in the contract.		Low	
	Security Concern/Implication If information regarding the level of support required from the provider and provisions for obtaining that support are not included in the recovery services contract, it is possible that technical support will be unavailable, thus slowing recovery or making recovery impossible. **Recommendation** We recommend that if an alternate site or other recovery service is used, the level of support required from the provider, and provisions for obtaining that support are included in the contract in order to ensure that the proper technical support is available during a test or an actual recovery.	**Manual Review Procedure** Review all recovery service contracts to ensure that an appropriate level of support is included. **Automated Tool Reference** N/A		

Detailed audit activities to be performed:

- Depending on the underlying legal system, verify that the service levels have been agreed in a binding manner. Obtain legal advice where appropriate.
- Review in detail that services and support will be delivered as an a priori part of the contract. Contractual remedies should be examined.

Refer to the results of a separate and independent audit of IT disaster recovery where such an audit has been performed. Alternatively, activities for this item should be conducted at a high level in order to avoid becoming encumbered with IT-related questions. Highlight any findings – such as the absence of appropriate IT disaster recovery plans complementing the business continuity management process – and issue recommendations accordingly.

		Risk	Workpaper References
6.4.16 Key contacts affiliated with the alternate processing site, and their phone numbers, are included in the business continuity plan.		Low	
Security Concern/Implication If key contacts affiliated with the alternate processing site and their phone numbers are not included in the business continuity plan, there is a potential that personnel will not know who to call at the alternate site in the event of a disaster. This could substantially delay a recovery effort. **Recommendation** We recommend that key contacts affiliated with the alternate processing site and their phone numbers are included in the business continuity plan in order to ensure their availability if a disaster were to occur.	**Manual Review Procedure** Review the business continuity plan to ensure that key contacts affiliated with the alternate processing site, and their phone numbers, are included. **Automated Tool Reference** N/A		

Detailed audit activities to be performed:

- Conduct test calls to numbers and fax numbers (in the latter case, measure and log response times).
- Check completeness and provider support levels / service levels.

			Risk	Workpaper References
	6.4.17 Physical access controls exist at the alternate facility (i.e., card keys, guard stations).		Low	
	Security Concern/Implication If appropriate physical access controls do not exist at the alternate facility, it is possible for unauthorised personnel to obtain access to the facility. This could lead to unauthorised people having access to confidential information. **Recommendation** We recommend that appropriate physical access controls exist at the alternate facility.	**Manual Review Procedure** Query management or the recovery site vendor as to what physical access controls exist at the alternate facility. **Automated Tool Reference** N/A		

Detailed audit activities to be performed:

- Refer to Security Architecture for detailed steps (link between IT security and physical security)
- Verify security measures and systems from perimeter inwards (building security, inside building, various sensors, central alarm system etc.). Consult experts where appropriate.
- Review security documentation and instructions to personnel.
- Review contracts with external security / guarding services.
- Include physical security in testing and maintenance of plans - check whether this has been done or is planned.

Physical access controls, both at the primary site and at an optional secondary site, are not the main target of BCM-related audit activities. Focus on verifying that the set of controls at the alternate site is at least equivalent to the level of security found under normal circumstances. Detailed audit steps should be avoided in order to minimize the time spent on this low-risk issue. It may be advisable to highlight weaknesses and include a recommendation for a full security audit which does not form part of the BCM review.

		Risk	Workpaper References
6.4.18 Procedures for enforcing access restrictions for the network environment at the recovery facility are in place and included in the business continuity plan.		Low	
Security Concern/Implication If procedures for enforcing access restrictions on the network environment at the recovery facility are not in place, there is a potential risk for an intruder to gain unauthorized access to confidential information. **Recommendation** We recommend that procedures for enforcing access restrictions on the network environment at the recovery facility are in place and included in the business continuity plan.	**Manual Review Procedure** Review the business continuity plan to ensure that procedures for enforcing access restrictions on the network environment at the recovery facility are in place and included in the plan document. **Automated Tool Reference** N/A		

Detailed audit activities to be performed:

- apply appropriate models of IT security
- depending upon the countries and locations involved, apply acknowledged standards such as BS 7799, BSI etc. to the environment under review
- compare security under normal ops with the instructions and access rights in alternative ops / recovery / restore phases

Logical access controls, both at the primary site and at an optional secondary site, are not the main target of BCM-related audit activities. Focus on verifying that the set of controls at the alternate site is at least equivalent to the level of security found under normal circumstances. Detailed audit steps should be avoided in order to minimize the time spent on this low-risk issue. It may be advisable to highlight weaknesses and include a recommendation for a full security audit which does not form part of the BCM review.

			Risk	Workpaper References
	6.4.19 Provisions for mail service at the primary site and the recovery site are included in the business continuity plan.		Low	
	Security Concern/Implication If provisions for mail service at the primary and recovery sites are not included in the business continuity plan, it is possible that important documents will not be delivered or mailed in an appropriate timeframe, which could lead to dissatisfied customers and a loss of business. **Recommendation** We recommend that provisions for mail service at the primary site and the recovery site are included in the business continuity plan.	**Manual Review Procedure** Review the business continuity plan to ensure that provisions for mail service at the recovery site are included in the document. **Automated Tool Reference** N/A		

		Risk	Workpaper References
6.4.20 Provisions for salvage, cleanup, and rebuilding of the primary site are included in the business continuity plan.		Medium	cf. strategy (rebuild, create new)
Security Concern/Implication If provisions for salvage, cleanup, and rebuilding of the primary site are not included in the business continuity plan, a return to normal business day functions at the primary site could be delayed. **Recommendation** We recommend that provisions for salvage, cleanup, and rebuilding of the primary site are included in the business continuity plan.	**Manual Review Procedure** Review the business continuity plan to ensure that provisions for salvage, cleanup and rebuilding of the primary site are included in the document. **Automated Tool Reference** N/A		

Detailed audit activities to be performed:

- Review the crisis management, recovery and restoration parts of business continuity planning. Check that initial and subsequent procedures for identifying salvageable equipment have been included.
- Verify that salvage procedures have been defined. These should not be in conflict with any superseding directives or orders issued by public authorities (*see Chapter 10*).
- Check that cleanup procedures have been included in the plans. These should be triggered by the official end of salvage activities. Cleanup should take place in accordance with any restrictions imposed by public authorities.
- Examine in detail selected restoration plans for the primary facility. Check that procedures for rebuilding have been included at a sufficient level of detail.

		Risk	Workpaper References
6.4.21 Provisions and procedures for manual operations, manual fallback solutions and related activities are included in the plan. Manual fallback procedures ensure that alternative operations can deliver at least 60 % of normal process performance.		**Low**	
Security Concern/Implication If manual operations have not been anticipated, it is possible that tool-based or technology-assisted alternative operations will not cover the residual risk. Depending upon the nature of the risk or disruptive event, manual operations may be the only viable recovery alternative. **Recommendation** Where there is a residual risk of failure intrinsic to technology-based solutions such as PC-based alternative operations, it is recommended that additional, manual procedures be introduced to ensure process performance at the desired level (60 %).	**Manual Review Procedure** Review technology-based procedures for alternative operations. Examine and critically assess manual workarounds, specifically forms, material in stock, etc. If manual work involves the use of legacy technology such as typewriter or telex, make sure that appropriate training has been held for users. **Automated Tool Reference** n/a		

Detailed audit activities to be performed:

- Review sample plans – at divisional, departmental, and group levels – and identify any manual workarounds or fallback procedures. Check that these provide an adequate level of service.
- Clarify the percentage (60 % minimum) agreed by senior management and BC personnel. It may be necessary to set a higher percentage if manual procedures are utilized to support a critical function or process.

- Review training plans and logs for selected manual procedures. Verify that all BC personnel and other staff involved in manual work have been appropriately trained. As a means of testing, visual inspection of manual equipment and observation of staff operating the equipment may be necessary.

The main difficulty with manual operations is that they complement, but do not normally replace, modern processes. In IT-supported environments in particular, the manual procedures used before may not be powerful enough to accomplish an adequate level of output. Processes that used to be performed using a telex are likely to have been made much more efficient by using fax and e-mail. It is unlikely that falling back onto the manual procedure of sending telexes will realistically solve the problem of achieving 60 % of process output. The backlog of falling back to telex should therefore be examined in detail.

In many instances, auditee organizations will state that "falling back to the old way of doing things" is seen as the cheapest and most logical solution to the problem of sustaining alternative operations. However, the auditor should investigate this very thoroughly. Given that the old technology or manual procedure has not been used for a number of years, employee knowledge and ability to perform according to manual concepts may not be guaranteed. Training and the demonstrated knowledge of all staff envisaged to carry out manual steps should be verified in detail. Similarly, but not necessarily as part of the BCM audit, the operation of old IT systems and programs might be subjected to these tests: are there enough staff who are knowledgeable enough to understand and operate old systems? Has this capability been tested and reinforced by appropriate training?

"Manual" in most of its connotations means "with your hands." Using electrical appliances or electronic equipment is not strictly manual in the sense of the word. If a manual procedure has been inserted as part of business continuity management, it will usually refer to paper and pencil (which is the only means of being entirely independent of power, networks, and technology). When semi-manual processes are defined in the business continuity management process, these should be examined with a view to establishing their technology dependence, i.e.: when a fax or telex machine is envisaged as part of "manual" operations, are power and connectivity guaranteed?

			Risk	Workpaper References
	6.4.22 The alternate processing facilities (i.e., hot site, duplicate data center) are appropriate, considering the MTD / MTA values.		Low	
	Security Concern/Implication If a recovery solution is chosen that takes 48 hours to implement, but the MTD / MTA is 24 hours, the critical business processes may not be recovered in a timely manner. This could lead to a slow or unsuccessful recovery effort. **Recommendation** We recommend that alternate processing facilities are appropriate, considering the MTD / MTA identified.	**Manual Review Procedure** Review the business continuity plan to determine what the MTD and MTA are for critical business processes. Compare these to the length of time it will take to implement the recovery strategy. Ensure that the alternate processing facilities will be adequate to support recovery within MTD or MTA. **Automated Tool Reference** N/A		

Detailed audit activities:

- Verify BIA results against strategy and detailed solutions in plans. Assess feasibility and adequacy of solutions.
- Apply methods of capacity planning / load balancing to determine whether processing facilities (alternate) are okay.
- Review in detail how infrastructural elements at the alternate site are planned and implemented. Compare with primary site facilities and assess adequacy.
- Consider alternative operations (in % performance of normal operations) and equate this with technical solutions implemented.

			Risk	Workpaper References
	6.4.23 Procedures for off-site storage and retrieval of vital paper records have been included in the BCM concept.		High	
	<u>Security Concern/Implication</u> If vital records have not been retained, or are unavailable in the event of a disaster, the organization may not be able to function normally or fulfil its contractual obligations towards business partners. <u>Recommendation</u> All vital records, such as original contracts, deeds, and other legally relevant material, should be stored off-site in a secure location. There should be a defined retrieval procedure for critical or disastrous events.	<u>Manual Review Procedure</u> Obtain information about the storage location and the extent of vital records stored there. <u>Automated Tool Reference</u> N/A		

Detailed audit activities to be performed:

- Obtain and review the list of records and written documents that are deemed "vital" to the functioning of the organization under normal circumstances.
- Distinguish between the levels of significance and criticality assigned to the material: some records may have legal significance, whereas others may be more important in a functional sense. Consult external experts where appropriate to find out about legally mandated storage or paper formats.
- Highlight any findings with regard to incomplete lists or misunderstandings of the term "vital." Recommend amendments to the list of vital records.
- Verify that the means and location for storing vital records are adequate. Some originals (such as old physical originals of deeds on land) may require special storage.
- Check the retrieval or reproduction mechanism utilized in the event of a crisis or disaster. Verify that vital original records are not put at risk by this. Recommend appropriate improvements.

			Risk	Workpaper References
	6.4.24 Third parties involved in the business continuity effort have business continuity plans that are adequate to satisfy the requirements and provide continuous delivery of products and services.		High	
	Security Concern/Implication If products and services envisaged in the business continuity plan are not delivered on time, recovery and restoration may be delayed, thus putting the organization at risk. **Recommendation** All products and services that are necessary for the recovery and restoration effort should be defined in the suppliers′ business continuity plans.	**Manual Review Procedure** Obtain appropriate assurance from the suppliers. **Automated Tool Reference** N/A		

Detailed audit activities to be performed:

- Take samples from lists of equipment to be supplied by third parties. Verify that the delivery of these materials is guaranteed by the supplier. This should not be restricted to a service level agreement; the supplier should be able to prove that product will be delivered on time.
- In a similar manner, check services that have been included as vital parts of the business continuity plan. These services should be guaranteed.
- Verify that both product and service delivery are an integral part of regular BCM testing (*see chapter 8*). Highlight any weaknesses documented in past test reports.
- Scrutinize the service level agreements (if any) to identify both the terms of delivery and the legal remedies specified. Consult external experts where appropriate.

The business continuity plans of all suppliers should clearly mirror the needs of the organization under review. Products and services mentioned in the business continuity plan are usually critical to the functioning of the organization in alternative operations. Any delay, non-delivery or other shortcomings would clearly endanger the recovery and restoration effort. It is therefore essential that all resources brought in from the outside in the event of a crisis or disaster must be delivered on time and in the manner specified in the business continuity plan.

		Risk	Workpaper References
6.5 Detailed Implementation			
6.5.1 Completion of detailed plans has been monitored and validated.		High	
Security Concern/Implication If detailed plans are incomplete or divergent in terms of content and level of detail, areas of inadequate planning may exist. This may render some plans dysfunctional in certain circumstances. **Recommendation** A monitoring and validation process should exist to verify completeness and level of detail for all continuity plans and documents.	**Manual Review Procedure** Obtain and review sample plans for accuracy, level of detail and conformance to overall planning guidelines. **Automated Tool Reference** Use software-based planning tools (where appropriate) to generate reports indicating missing plans or information.		

Detailed audit activities to be performed:

- Take samples of plans and compare them against the general standard of planning and documentation. Highlight any business processes or functions not covered by detailed plans and instructions. Identify and report high-risk areas where planning or implementation of continuity-related measures appears to be inadequate or incomplete.
- Where software for BC planning and management is being used (also refer to Section 12 of this program), generate reports highlighting incomplete or inaccurately filled parts of the planning database.
- Review the monitoring process and the validation mechanisms applied to detailed business continuity plans. Check if all plans have been verified by business continuity personnel, technical experts and department heads.

			Risk	Workpaper References
	6.5.2 A planning glossary has been prepared providing a uniform set of definitions for detailed plan implementation.		Mediu m	
	Security Concern/Implication If no common glossary of words, notions and definitions exists, misunderstandings may result which lead to misguided planning and dysfunctional plans. Re-planning and rework may be necessary. **Recommendation** A common, uniform glossary for all terms, definitions and words relating to BCM should be prepared and distributed prior to commencing detailed planning and BCM deployment.	**Manual Review Procedure** Obtain and review the glossary or glossaries of BCM terms. Verify that multilingual glossaries are correct and sufficiently detailed. **Automated Tool Reference** N/A		

Detailed audit activities to be performed:

- Obtain copies of glossaries used in the project initiation and management phase. Check against common wordings found in standards and other BCM-related works.
- Use dictionaries and external expertise to verify that all English words are translated correctly into other language versions of the glossary. Highlight discrepancies and cultural problems arising from translation errors.
- Conduct interviews with local management to ascertain the adequacy and usability of the glossary / -ies. Compile a list of desirable or necessary improvements.
- It is important that all countries and nationalities which form part of overall BCM deployment "speak the same language." The glossary, as understood by local managers and BCM experts, may turn out to be a central document in terms of alignment with the corporate BCM philosophy. Although this item is classified as "medium" risk,, errors and omissions in the glossary may lead to misunderstandings and enormous amounts of rework on detailed plans. This should be borne in mind when validating the terminology used on an enterprise-wide basis.

	Risk	Workpaper References
6.5.3 All plans are in compliance with planning guidelines. The contents of plans do not diverge significantly from these guidelines	High	
Security Concern/Implication If individual plans do not follow the guidelines and the standardized format set by the business continuity management framework, there is a risk of misunderstandings, incorrect actions described in plans, and dysfunctional plans. **Recommendation** All plans should follow the planning guidelines as set down in the BCM framework. Significant divergence should be identified, and plans should be updated to an adequate standard.		
Manual Review Procedure Take appropriate samples of detailed plans at various levels of the organization. Compare plan contents with the guidelines defined in the BCM framework. Highlight any discrepancies or inconsistencies. **Automated Tool Reference** N/A		

Detailed audit activities to be performed:

- Obtain a representative sample of detailed plans, taken from various areas and departments of the auditee entities or organization(s).
- Review the plans. Highlight any inconsistencies and divergence from the approved BCM standard set prior to commencing detailed planning. Where such divergence exists, take additional samples of plans to verify if a fundamental misunderstanding with regard to planning has occurred.

6.6 Plan Distribution and Control		Risk	Workpaper References
6.6.1 The plan distribution list adequately reflects the "need to know" principle as defined by senior management.		High	
Security Concern/Implication If plans are unrestricted in terms of distribution, there is a risk of inadvertent disclosure. Given the often sensitive nature of information contained in business continuity plans, this may pose an increased risk to the organization. Indiscriminate distribution of various levels of plans may also lead to confusion at lower organizational levels whose main concern is with their own plans. **Recommendation** The plan distribution list should be defined in accordance with the "need to know" principle and the guidelines set by senior management. Too much information for lower levels of the BC team structure should be avoided, as should be too little information at middle and higher echelons of management.	**Manual Review Procedure** Review the plan distribution and question selected items. Ensure that the "need to know" principle has been adhered to and that management are aware of the contents of the distribution list. Compare the list with security arrangements and security guidelines. **Automated Tool Reference** Typical security-related toolsets, data classifications, access matrices		

Detailed audit activities to be performed:

- Review the distribution list and the levels of planning accessible to persons within the organization. Compare their access rights to BC plans to other typical access rights such as IT security or physical security. Verify that individuals having access to BC plans are given all pertinent information that is required to perform their assigned tasks and duties in terms of BCM.
- Highlight any discrepancies (too much or too little BC-related information) and discuss possible reasons with the individuals concerned and their supervisors. Compare findings with the "need to know" principle and organizational communications policies as defined by senior management. Report any persistent discrepancies and elaborate on possible consequences for the functioning of BC plans and BC teams.

	Risk	Workpaper References
6.6.2 Procedures for regular distribution of updates have been included. A version control mechanism is in operation.	High	
Security Concern/Implication If updates are not controlled by an appropriate procedure, different versions of plans may be in existence throughout the organization. If no version control is established, there may be ambiguity as to the current version and its dissemination throughout the organization. Both these risks increase the probability of out-of-date plans. **Recommendation** The distribution of updates should be controlled by a procedure that is included with the distribution list. Versions, numbering and version control should be built into the distribution list in a simple and transparent manner.		
Manual Review Procedure Randomly check versions of detailed plans. Review the distribution list and verify that there is a mechanism for controlling versions, version numbering and feedback by the recipient. **Automated Tool Reference** Where BC software is being used, refer to the relevant chapter in this section.		

Detailed audit activities to be performed:

- Obtain random samples of plans, including those obtained in the course of covering *chapter 6.5* of this program. Check version numbers and version consistency. Highlight any discrepancies.
- Check that the distribution list has attachments or text sections describing the update / update distribution procedure. Verify that a feedback loop has been built in to ensure that the recipient is aware of the version change and plan update.

		Risk	Workpaper References
6.6.3 The plan distribution and control procedures cover third parties in an adequate manner.		High	
Security Concern/Implication If third parties are not included in the distribution and control procedures, their involvement in BC may be at risk. They may not be able to perform their assigned tasks and duties in the event of a crisis or emergency. **Recommendation** Third parties who have been assigned tasks and duties as part of business continuity management should be included in the distribution and control procedures for BC plans. The extent of plan information disclosed to them should be in line with "need to know" guidelines.	**Manual Review Procedure** Recursively, as described in *6.6.1* and *6.6.2* **Automated Tool Reference** N/A		

Detailed audit activities to be performed:

- as in *6.6.1* and *6.6.2*.
- Verify that third parties are included in plan distribution and version control to a limited extent. Their knowledge should be defined by the "need to know" principle and by the nature and extent of their involvement in business continuity management. For instance, a vital hardware and network component supplier may need to be given more extensive access to plans than a finance department (internal) employee. Highlight any misunderstandings, discrepancies and issues that may lead to dysfunctional plans. Make sure that "vital" or "critical" suppliers are not excluded from information that they need to perform their assigned contractual duties.

Overview Chapter 6 Audit Items

Number	Preliminary Rating Assigned by Auditor (from 1 to 5, 1 = absent or inadequate, 5 = comprehensive and adequate), and auditor's comment	Risk
6.2.1		High
6.2.2		High
6.2.3		High
6.3.1		High
6.4.1		High
6.4.23		High
6.4.24		High
6.5.1		High
6.5.3		High
6.6.1		High
6.6.2		High
6.6.3		High

6.1.1		Medium
6.2.4		Medium
6.3.2		Medium
6.3.3		Medium
6.4.3		Medium
6.4.6		Medium
6.4.7		Medium
6.4.10		Medium
6.4.12		Medium
6.4.20		Medium
6.5.2		Medium

6.1.2		Low
6.1.3		Low
6.4.2		Low
6.4.4		Low
6.4.5		Low
6.4.8		Low
6.4.9		Low
6.4.11		Low
6.4.13		Low
6.4.14		Low
6.4.15		Low
6.4.16		Low
6.4.17		Low
6.4.18		Low
6.4.19		Low
6.4.21		Low
6.4.22		Low

<table>
<tr><td colspan="3"><h1>7 Training and Awareness</h1></td><td colspan="2"></td></tr>
<tr><td colspan="3"><h2>7.1 Business Continuity Awareness</h2></td><td>Risk</td><td>Workpaper References</td></tr>
<tr><td></td><td colspan="2">7.1.1 BCM is supported by the industry, relevant associations and within the company. The legal, regulatory and cultural background reinforces BCM awareness.</td><td>Low</td><td></td></tr>
<tr><td></td><td><u>Security Concern/Implication</u>
If the economic and cultural environment of the company does not support BCM, there may be a discrepancy between internal and external perceptions. This may lead to a lack of credibility of the company as seen from the outside.

<u>Recommendation</u>
Internal training and BCM policies should be designed to adequately reflect the surrounding economic and cultural situation. Where a low priority environment for BCM is in evidence, the internal awareness program should reinforce the corporate vision of BCM</td><td><u>Manual Review Procedure</u>
Obtain an understanding of the cultural and economic background that may influence BCM in the country or region being audited. Compare corporate (internal) perception of BCM with the general understanding throughout the industry.

<u>Automated Tool Reference</u></td><td></td><td></td></tr>
</table>

Detailed audit activities to be performed:

- Refer to relevant chapters of this audit program to establish the legal and regulatory background of the organization under review. Conduct interviews with opinion leaders or industry association representatives to gain an understanding of the cultural situation in the country or region. Use the network of auditors or other experienced practitioners to gather secondary opinions on BC culture and awareness.
- Compare the organization's stance on BC with the prevailing cultural opinion or social trends. If there are marked differences, review the steps taken by the auditee entity to create an internal BC culture and "corporate history." Where the organizational culture appears to be weak in comparison with external thinking, highlight and report this fact. Where internal BC thinking is comparatively strong, this should not be seen as an audit finding.

NB: this audit item is often extremely important, but difficult to verify using hard facts. Many organizations are facing an internal culture imported from abroad (for instance, UK or US subsidiaries in other regions of the world), whereas local firms may be strongly influenced by the prevalent mode of thinking in the country. With regard to overall BC efficiency and awareness, the programs and initiatives taken by the auditee entity should show a clear and profound understanding of the cultural environment they are designed for.

7.2 BCM Training and Awareness		Risk	Workpaper References
7.2.1 BCM personnel are trained and educated on a regular basis. Internal and external courses or knowledge management activities are in place. Formal training plans have been documented for BCM personnel.		Low	
Security Concern/Implication A lack of training on an individual or collective basis may lead to weaknesses in planning and managing business continuity. **Recommendation** We recommend the definition of formal training and skills development plans for BCM personnel. These should be complemented by other learning activities initiated by BCM staff; such activities should be actively encouraged by senior management.	**Manual Review Procedure** Obtain and review training plans (internal, for BCM personnel). Review documentation of training and other learning activities during the past year. **Automated Tool Reference**		

Detailed audit activities to be performed:

- Obtain training plans (as agreed in appraisal sessions or other formal meetings) and compare with actual training done in the past year. Note discrepancies and reasons.
- Obtain list of informal - voluntary - learning activities initiated by BCM personnel and assess quality. Review support and (where applicable) corporate policy for financing such activities.
- Review future training plans and mechanisms for definition of these plans (superior's decision, interactive discussion, employee's decision, budget constrained selection etc.).
- Assess adequacy of training plans over a time horizon of 1 year, 3 years, 5 years.

		Risk	Workpaper References
7.2.2 Other staff are trained and educated about BCM on a regular basis. These educational sessions are compulsory and supported by senior management. Formal induction procedures exist for new recruits in order to convey the BCM message to them.		**Medium**	
Security Concern/Implication If employees are not trained and educated in BCM matters, reactive capability of the client organization as a whole may be compromised. Technical and habitual obsolescence of BCM thinking and emergency behavior may lead to extremely dangerous situations during a business interruption. **Recommendation** Employees should be trained in BCM matters on a regular basis. New recruits should receive adequate information with regard to BCM procedures and duties.	**Manual Review Procedure** Review training plans and course schedule for the past year. Collect and assess materials handed to new employees. **Automated Tool Reference**		

Overview Chapter 7 Audit Items

Number	Preliminary Rating Assigned by Auditor (from 1 to 5, 1 = absent or inadequate, 5 = comprehensive and adequate), and auditor's comment	Risk
7.2.2		Medium
7.1.1		Low
7.1.2		Low

8 Maintenance and Exercise		
8.1 Plan Testing	**Risk**	**Workpaper References**
8.1.1 All test results and plan deficiencies are documented during and after the test of the business continuity plan.	High	Rothstein (ed.): Disaster Recovery Testing: Exercising Your Contingency Plan, Rothstein Associates 1995; ISBN 0-9641648-0-9.
Security Concern/Implication If test results and plan deficiencies are not documented during and after the test of the business continuity plan, changes to be made to the plan which were discovered during testing may be forgotten. A plan which is not current and properly updated can lead to an inefficient and ineffective recovery. **Recommendation** We recommend that all test results and plan deficiencies are documented during and after the test of the business continuity plan. **Manual Review Procedure** Review the test results from the most recent test. Cross-reference the comments from the test to the business continuity plan to ensure that all items noted resulted in a proper update being made to the plan. **Automated Tool Reference** N/A		

Detailed audit activities to be performed:

- Obtain the test documentation for the most recent tests performed. Review test results and recommendations for improvement. Check that the plans have been updated accordingly.

		Risk	Workpaper References
8.1.2 Both scheduled and unscheduled tests of the business continuity plan are performed.		Medium	Rothstein (1995), as above
Security Concern/Implication If only scheduled tests of the plan are performed, personnel may have time to "prepare" for the test by ensuring that certain information is available to them during the test. This would not be an accurate reflection of what could potentially occur in a disaster. **Recommendation** We recommend that both scheduled and unscheduled tests of the business continuity plan are performed.	**Manual Review Procedure** Review test results to determine the scope of each of the tests. Ensure that both scheduled and unscheduled tests have been performed. **Automated Tool Reference** N/A		

Detailed audit activities to be performed:

- Compare the scope statements for a selected series of recent tests. Verify that a test schedule exists. Highlight the absence of unscheduled tests.
- In certain test schedules spanning several years, it is possible that unscheduled tests are not envisaged for the first year or second year following completion of the first version of business continuity management. When reviewing such programs, make sure that the development of more elaborate tests and BC exercises follows a clear pattern that is in line with organizational BC maturity.

		Risk	Workpaper References
8.1.3 Detailed test plans, including test objectives, scope, and timing exist for every test performed.		Medium	
Security Concern/Implication If test plans are not clear and do not include pertinent information such as test objectives, scope, and timing, the test may not accomplish all the predetermined objectives. If the test does not accomplish what it set out to do, information in the plan that is outdated or incorrect may not be discovered. **Recommendation** We recommend that for each test that is conducted, detailed test plans exist so that the scope of the test is clearly laid out, and the test accomplishes all the pre-determined objectives.	**Manual Review Procedure** Query management as to whether test plans were developed for past tests, and if one is currently being developed for the next test. Review the test plan(s) to ensure that scope, objectives, and timing are included. **Automated Tool Reference** N/A		

Detailed audit activities to be performed:

- Verify that there are reasonable test cycles in terms of a) target, b) scope, c) proportion of scheduled vs. unannounced tests.
- Check test scope and realism of test objectives.
- Review all tests to date (not necessarily in detail) and collect evidence showing that there is no particular bias in testing, no blind spots.

		Risk	Workpaper References
8.1.4 Tests of the business continuity plan are performed after there is a major change in the company environment.		Medium	
Security Concern/Implication If tests of the business continuity plan are not performed after there is a major change in the company environment, it is possible that the plan will not accurately reflect recovery for the new environment, and recovery could be insufficient or unsuccessful. **Recommendation** We recommend that tests of the business continuity plan are performed after there is a major change in the company environment.	**Manual Review Procedure** Query management as to whether tests of the business continuity plan are performed after there is a major change in the company environment. **Automated Tool Reference** N/A		

		Risk	Workpaper References
8.1.5 The business continuity plan is tested on a regular and frequent basis, and at least once per year.		Medium	
Security Concern/Implication If the business continuity plan is not tested regularly, personnel may not be familiar with the roles they are to perform, and the plan may contain discrepancies in it. This could lead to a slow or ineffective recovery effort. **Recommendation** We recommend that the business continuity plan be tested regularly. If possible, we recommend that the plan be tested at least on a semi-annual basis.	**Manual Review Procedure** Query management as to how often the business continuity plan is tested and what is included in each test. Review any test schedules to ensure that the plan is being tested frequently. **Automated Tool Reference** N/A		

		Risk	Workpaper References
8.1.6 The test scope, addresses a range of scenarios during both peak and off-peak processing times.		**Medium**	
Security Concern/Implication If the test scope does not address a range of potential disruptions during both peak and off-peak processing times, personnel may only be familiar with only one particular scenario. This could lead to confusion regarding various recovery procedures. **Recommendation** We recommend that the test scope addresses a range of scenarios during both peak and non-peak processing times.	**Manual Review Procedure** Review past test results and test plans for future tests to ensure that the test scope addresses a range of scenarios during both peak and off-peak processing times. **Automated Tool Reference** N/A		

Detailed audit activities to be performed:

- Test scenarios and related hypothetical situations should cover effects rather than causes, e.g. Scenario 1: loss of several floors; Scenario 2: loss of 3 buildings; etc. Test scenarios may be event-based, but any test objective should be related to effects, not causes.
- Check whether events are viewed in terms of their consequences. Test situations should clearly prevent participants from enacting disasters on a global scale, i.e., beyond the immediate corporate environment.

<table>
<tr><td colspan="3"></td><td colspan="2"></td></tr>
<tr><td colspan="3">8.2 Plan Maintenance</td><td>Risk</td><td>Workpaper References</td></tr>
<tr><td></td><td colspan="2">8.2.1 Business continuity plan versions are numbered to avoid accidental use of an old plan version.</td><td>Low</td><td></td></tr>
<tr><td></td><td><u>Security Concern/Implication</u>
If an incorrect version of the business continuity plan is used during a recovery, it is possible that inaccurate information will be used, thus hindering the recovery effort.

<u>Recommendation</u>
We recommend consecutively numbering each new version of the business continuity plan. All old versions of the plan should be discarded as soon as a new version is distributed in order to avoid confusion.</td><td><u>Manual Review Procedure</u>
Review the business continuity plan to ensure that it is numbered and/or dated. Query management as to whether there is a policy in place to ensure that all old versions of the plan are discarded immediately upon receipt of a new version.

<u>Automated Tool Reference</u>

version control mechanisms (electronic or manual)</td><td></td><td></td></tr>
</table>

Detailed audit activities to be performed:

- Examine the numbering / version control mechanism in operation. When using electronic BC tools or document management systems, familiarize with the relevant parts of these tools.
- Refer to *chapter 6* of this program to check if the version control and numbering mechanism has been included in plan distribution and control procedures.

		Risk	Workpaper References
8.2.2 Documented procedures exist to keep the business continuity plan current following changes made to the organization (new systems, organizational changes, etc.).		**Low**	
Security Concern/Implication If no documented procedures exist to update the business continuity plan following organizational changes, it is possible that the plan will not be current which could cause an unsuccessful recovery effort. **Recommendation** We recommend instituting a procedure to keep the business continuity plan current following changes made to the organization.	**Manual Review Procedure** Query management as to any policies that are in place which require business continuity plan updates following any organizational changes. **Automated Tool Reference** N/A		

Detailed audit activities to be performed:

- Review the policies and documents describing plan maintenance and tasks following an organizational or technical change.

Where a single individual is responsible for maintaining BC plans, refer to *question 8.2.3,* below.

		Risk	Workpaper References
8.2.3 Responsibility for the maintenance of the business continuity plan is assigned to a specific individual or department.		High	
Security Concern/Implication Having multiple or no parties responsible for the business continuity plan will result in a lack of ownership, and an out-of-date plan. **Recommendation** We recommend assigning a specific individual or department to be responsible for ensuring that the business continuity plan is accurate and up-to-date.	**Manual Review Procedure** Query management as to whether there is an individual or specific department who takes responsibility for the business continuity plan. Speak with this individual and ensure that this person is keeping the plan current. **Automated Tool Reference** N/A		

Detailed audit activities to be performed:

- Conduct an interview with the person or department head responsible for maintaining BC plans. Note the procedures and frequency of maintenance runs. Verify that the maintenance activities performed during the year follow the documented guidelines.

<table>
<tr><td colspan="3"></td><td>Risk</td><td>Workpaper References</td></tr>
<tr><td></td><td colspan="2">8.2.4 The business continuity plan document is maintained both on-site and in an off-site location.</td><td>Medium</td><td></td></tr>
<tr><td></td><td><u>Security Concern/Implication</u>
Storing the business continuity plan document in only one location could result in the document being unavailable at the time of a disruption.

<u>Recommendation</u>
We recommend storing the business continuity plan in multiple locations. This helps to ensure that there is access to the plan at all times.</td><td><u>Manual Review Procedure</u>
Query management and personnel as to where they keep their business continuity plan documents. Ascertain whether at least one copy of the plan is maintained in a secure, off-site location.

<u>Automated Tool Reference</u>
N/A</td><td></td><td></td></tr>
</table>

		Risk	Workpaper References
8.2.5 The business continuity plan is updated at least semi-annually.		**Medium**	
Security Concern/Implication Business continuity plans that are not updated at least semi-annually may not be effective as a result of personnel changes, phone number changes, hardware and software changes, and organizational changes that may affect the plan. **Recommendation** We recommend instituting a policy that the business continuity plan is updated a minimum of twice a year. We recommend assigning dates to have this task completed by.	**Manual Review Procedure** Review the plan and verify that all information and contacts are accurate. Query management as to the date the last update was performed and when the next one is scheduled to be performed. **Automated Tool Reference** N/A		

Overview Chapter 8 Audit Items

Number	Preliminary Rating Assigned by Auditor (from 1 to 5, 1 = absent or inadequate, 5 = comprehensive and adequate), and auditor's comment	Risk
8.1.1		High
8.2.3		High
8.1.2		Medium
8.1.3		Medium
8.1.4		Medium
8.1.5		Medium
8.1.6		Medium
8.2.4		Medium
8.2.5		Medium
8.2.1		Low
8.2.2		Low

9 Public Relations and Communications			
9.1 Public Relations		**Risk**	**Workpaper References**
9.1.1 All relevant components of a Public Relations concept have been included in the business continuity plan.		**High**	
Security Concern/Implication If the public relations concept does not cover all relevant areas and business processes, open issues and questions may arise which are not adequately addressed. This may lead to discrepancies, misunderstandings and unfounded suspicions voiced by stakeholders. **Recommendation** The P.R. concept should include all relevant areas and business processes which might be affected by a disruption or disaster. P.R. components should anticipate and pre-empt the major interests that stakeholders may have.	**Manual Review Procedure** Review the public relations framework and assess its adequacy in terms of anticipated stakeholder interests. **Automated Tool Reference**		

Detailed audit activities to be performed:

- Interview the P.R. department and obtain information on standard procedures as well as the general framework for public relations activities. Verify that crisis communications are perceived and planned as a separate and distinct mode of communications.
- Review the P.R. planning components related to business continuity, disruptive and disastrous events. Check for completeness and plausibility. Consult experts where necessary.
- Verify that senior management supports the P.R. concept and has approved the contents.

			Risk	Workpaper References
	9.1.2 External liaison points have been defined and are documented in the business continuity plan.		High	
	Security Concern/Implication If external liaison points are not defined, interaction between corporate management and external agencies may be impeded. Delays, inconsistencies and misunderstandings are likely to cause negative public perception and a loss of initiative for corporate business continuity management. **Recommendation** External liaison points should be established with all relevant agencies (e.g. emergency services, civil defense and local authorities) to ensure seamless coordination of the overall P. R. effort.	**Manual Review Procedure** Review the documentation for external liaison. Check if all relevant external agencies have been included in the plan. Verify that all plans are synchronized with communications strategies (see *section 10* of this program). **Automated Tool Reference**		

Detailed audit activities to be performed:

- Obtain and review individual documentation of external links with emergency services, civil defense, etc. Check for completeness, plausibility and verify that the individual plans have been agreed upon by both parties.
- Verify that corporate liaison plans and procedures are aligned with standard procedures used by the agency / authority reviewed. Highlight any discrepancies and possible area of dysfunctional interaction.

	Risk	Workpaper References
9.1.3 Stakeholders and interest groups have been identified and unambiguously described in the business continuity plan.	High	
Security Concern/Implication If stakeholders and interest groups have not been identified and integrated, the Public Relations concept may not address their requirements and interests. Individual stakeholders or groups may object to parts of the P.R. plan and therefore not cooperate in the event of a disaster or disruption. **Recommendation** All stakeholders and interest groups should be identified and included in the public relations plan. Appropriate steps should be taken to consider all individuals and groups which may be impacted by disruptive or disastrous events, and to include them in the P. R. effort. **Manual Review Procedure** Review the P. R. concept and verify that all stakeholders have been included in the planning process and the plan execution phase. **Automated Tool Reference**		

Detailed audit activities to be performed:

- Obtain a list of stakeholders as identified in the business continuity plan. Check for completeness, considering internal as well as external stakeholders who might have an interest in a disruptive / disastrous scenario.
- If necessary, interview stakeholders on a sample basis to ascertain the validity of the P. R. concept and stakeholder integration. Check if legitimate interests of individual stakeholders and / or groups have been considered in line with senior management directives and corporate policy.

9.2 Crisis Communications		Risk	Workpaper References
9.2.1 Procedures for handling press releases and public relations, including a generic media statement are included in the business continuity plan.		Low	
Security Concern/Implication If detailed procedures for handling press releases and public relations are not included in the business continuity plan, there is a potential for a negative or incorrect statement of the situation. **Recommendation** We recommend that procedures for handling press releases and public relations, including a standard media statement, are included in the business continuity plan to avoid a misunderstanding of the situation by the public.	**Manual Review Procedure** Review the business continuity plan to ensure that procedures for handling press releases and public relations, including a generic media statement, are included in the plan. Discuss with management whether there is a particular person who will be in charge of any interaction with the press and general public. **Automated Tool Reference** N/A		

Detailed audit activities to be performed:

- Verify whether planning addresses PR and press work in all phases of an incident (refer to disaster timeline)
- Check language issues and deployment of standardized PR strategy in different countries / cultures. Consult experts where necessary.
- Check if person(s) responsible has / have been trained.

		Risk	Workpaper References
9.2.2 Other crisis communications strategies have been included in the business continuity plan.		Low	
Security Concern/Implication If communications with internal or external stakeholders during a crisis have not been defined and clearly stated in the business continuity plan, there is a risk of negative or incorrect statements regarding the situation. This may result in increased stakeholder pressure and liability issues. **Recommendation** Communications with all internal and external stakeholders should be defined and clearly stated in the business continuity plan. The communications strategies used should be targeted and consistent.	**Manual Review Procedure** Review the business continuity plan to verify that clear strategies for communicating with internal and external stakeholders exist. Test senior management awareness with regard to these communications issues. **Automated Tool Reference** N/A		

Detailed audit activities to be performed:

- Verify whether planning addresses communications in all phases of an incident (refer to disaster timeline), covering all possible stakeholders.
- Check language issues and deployment of the communications strategy in different countries / cultures. Consult experts where necessary.
- Check if person(s) responsible has / have been trained.

Overview Chapter 9 Audit Items

Number	Preliminary Rating Assigned by Auditor (from 1 to 5, 1 = absent or inadequate, 5 = comprehensive and adequate), and auditor´s comment	Risk
9.1.1		High
9.1.2		High
9.1.3		High

9.2.1		Low
9.2.2		Low

10 Coordination with Public Authorities			
10. 1 Regulatory Framework		**Risk**	**Workpaper References**
10.1.1 The legal background for business continuity management has been thoroughly researched and used as a basis for planning.		**High**	
Security Concern/Implication If the legal background of business continuity management has not been adequately researched, plans may not be in compliance with the law or specific regulations, leading to liability and litigation risks. **Recommendation** The legal background for BCM should be researched, using specialist expertise where appropriate. All relevant Acts, statutes and other legal provisions should be known prior to commencing BC planning.	**Manual Review Procedure** Review the documentation on legal provisions governing BCM. Consult experts where appropriate. **Automated Tool Reference** Legal services, online databases		

Detailed audit activities to be performed:

- Referring to the particular country or geographical region in which the auditee organization is situated, obtain an overview of all legal issues that may influence business continuity management. These may be Acts, individual regulations, or other state-driven initiatives.
- Compile a list of requirements for BCM mandated by the legal background. Consult legal advisors where appropriate, and specifically where considerable doubt may exist as to the applicability of certain laws or regulations. Include the auditee organization's legal department to find out about their understanding of BCM.
- Verify that all legal aspects have been considered in defining the BCM framework. Highlight any doubtful areas and planning inconsistencies. Focus on typical areas where contradictory legal and factual situations may exist: fire protection, technical infrastructure etc. These may be indicative of the overall level of awareness.

		Risk	Workpaper References
10.1.2 Relevant industry standards and other pertinent regulations have been identified and used as a basis for planning.		High	
<u>Security Concern/Implication</u> If industry standards and regulations have not been adhered to in preparing the plans, there is an increased liability and litigation risk for the organization. Non-compliance may also mean that plans are deficient in terms of their functioning in the event of a crisis or disaster. **<u>Recommendation</u>** All industry regulations and recognized standards should be taken into account and adhered to when preparing business continuity plans.	**<u>Manual Review Procedure</u>** Obtain all standards and regulations from official sources. Perform a compliance review by examining selected plans. Highlight any areas of non-compliance. **<u>Automated Tool Reference</u>** Web search engines, government and industry association online libraries		

Detailed audit activities to be performed:

- Compile a list of relevant industry standards and industry regulations for the country or region in which the entity under review is located. Refer to *chapter 11* of this program for guidance. Consult experts where appropriate.
- Obtain and review the documentation relating to BCM and the incorporation of standards into the planning and management process.
- Report any findings that indicate non-compliance with standards and regulations. Consider compensating controls.

10.2 Coordination with Disaster Recovery and Business Continuity Agencies		Risk	Workpaper References
10.2.1 Statutory / official agencies have been identified and included in the process of business continuity planning.		High	
Security Concern/Implication If disaster recovery and business continuity or emergency management agencies have not been included in the planning process, conflicts between agency objectives and objectives pursued by the organization under review may occur. In the event of a crisis or disaster this may lead to confusion and conflicting interests. **Recommendation** All statutory or official agencies for disaster recovery, business continuity management and emergency management should be included in the business continuity management process.	**Manual Review Procedure** Identify relevant agencies for the country or region in question. Verify that BCM has included these in the scoping, planning and management processes. **Automated Tool Reference** N/A		

		Risk	Workpaper References
10.2.2 Liaison procedures with DR and BC agencies have been established and documented in the business continuity plan.		**Medium**	
Security Concern/Implication If there is no liaison with relevant DR and BC agencies in the location where the organization is situated, misunderstandings and counter-productive efforts may result. **Recommendation** Liaison procedures with relevant local DR and BC agencies should be established and documented in the business continuity plan	**Manual Review Procedure** Obtain and review correspondence and agreements with DR and BC agencies. **Automated Tool Reference** N/A		

			Risk	Workpaper References
	10.2.3 Coordination procedures for local authorities have been defined and documented.		Medium	
	Security Concern/Implication If no coordination mechanisms exist, interaction with local authorities in the event of a crisis or disaster may be deficient. As a result, contradictory action may be taken either by the company affected or by the local authorities having to deal with the "big picture." This may result in confusion, delay and an adverse relationship between corporate interests and public interest as defined and communicated by the authorities in question. **Recommendation** A comprehensive and approved set of coordination procedures for local authorities should be defined and documented.	**Manual Review Procedure** Review documents and procedural items pertaining to actions or reports between the corporate entity audited and local authorities. Request and examine documents showing agreement and approval between the parties. **Automated Tool Reference** N/A		

Detailed audit activities to be performed:

- Depending on the geographic region of the audit, assess the overall structure of local authorities. If necessary, obtain independent intelligence by consulting with government departments or seek the assistance of a local lawyer. Ensure that you are familiar with the structure, functioning and level of authority for each department or unit that may influence the recovery and restoration process.
- Verify that all possible concerns have been addressed. Conduct a detailed review of minutes, memos and other documents that may be used as evidence in the aftermath of a crisis. Ascertain that negotiations have taken place and that the final policy / set of procedures have been formally agreed upon by all parties involved.
- Verify that local authorities have been invited / have agreed to participate in BCM testing and maintenance (cf. *Section 8*).

		Risk	Workpaper References
10.2.4 Coordination procedures for police interaction have been defined and documented.		High	
Security Concern/Implication If no coordination procedures have been agreed with police forces, the corporate entity under review may not be able to initiate emergency response, recovery and restoration activities. There is a significant risk that police forces involved may impose strict discipline with regard to any action taken in the event of a crisis or disaster. **Recommendation** It is strongly recommended that coordination mechanisms with police forces be discussed, defined and mutually agreed upon. Issues and concerns raised by police officers should be balanced with corporate interests to avoid conflicts and ensure a clear line of action in the event of a crisis or disaster.	**Manual Review Procedure** Obtain and review all documentation relating to police coordination. Request and review relevant memos, minutes and approvals to ascertain whether mutual agreement has been reached between the parties. **Automated Tool Reference** N/A		

Detailed audit activities to be performed:

- Depending on the geographic location of the audit, obtain information as to the structure of local police forces. If necessary, consult experts or obtain independent intelligence (e.g., local lawyers). Ensure that you are familiar with the structure, functioning and level of authority for each department or unit that may influence the recovery and restoration process.
- Verify that coordination mechanisms have been discussed and negotiated with the police. Collect documentary evidence to show that the police have agreed to the course of action suggested by the entity audited. Note any apparent or implied conflicts of interest. Where appropriate, refer to *section 5* of this audit program to ensure that relevant scenarios with police involvement have been addressed and are properly coordinated.
- If appropriate, highlight areas that may not be accessible to the auditor, such as confidential arrangements for executive protection or classified material. Ensure that responsibility for these audit items (which may be beyond the control of the entity audited) rests with the police and that this fact has been documented.

			Risk	Workpaper References
	10.2.5 Coordination procedures for fire brigade interaction have been defined and documented.		**High**	
	Security Concern/Implication If no coordination mechanism with local or regional fire brigade command have been defined, corporate business continuity plans may be superseded by actions and instructions given by fire brigade officials during a crisis or disaster. Conflicts of interest may arise between fire brigade and the company. There is a risk that corporate interests may be endangered by official crisis management activities if no clear policies and procedures have been agreed beforehand. **Recommendation** A comprehensive, documented and mutually agreed coordination mechanism for fire brigade interaction should be introduced. If active or retired firemen are working within the company, these should be utilised as a "natural extension" of (external) professional fire brigade units.	**Manual Review Procedure** Obtain and review all documentation relating to fire brigade coordination. Request and review relevant memos, minutes and approvals to ascertain whether mutual agreement has been reached between the parties. **Automated Tool Reference** N/A		

Detailed audit activities to be performed:

- Depending on the geographic location of the audit, obtain information as to the structure of local fire brigade units. If necessary, consult experts or obtain independent intelligence (e.g. local government). Ensure that you are familiar with the structure, functioning and level of authority for each department or unit that may influence the recovery and restoration process.
- Document any relationships between the professional fire brigade and corporate personnel who are active or retired firemen. If intra-organizational experience exists, verify that it has been documented and listed in an appropriate manner. Active or retired firemen within the entity audited should be communicated to the local fire brigade units covering any crisis or disaster.
- Verify that coordination mechanisms have been discussed and negotiated with the fire brigade. Collect documentary evidence to show that the fire brigade units have agreed to the course of action suggested by the entity audited. Note any apparent or implied conflicts of interest. Where appropriate, refer to *section 5* of this audit programme to ensure that relevant scenarios with fire brigade involvement have been addressed and are properly coordinated.
- Verify if firemen within the organization have been included in the coordination effort, preferably as a "natural extension" of professional fire brigade units.
- Examine damage assessment mechanisms and other procedures for securing evidence in the event of a crisis or disaster. Bear in mind that collateral damage caused by fire fighting activities may be extensive, and that appropriate steps should be taken to substantiate such damage, if it cannot be avoided.
- Refer to the other sections of this audit programme to ensure that the relationship between BCM leadership, other departments and the fire brigade has been defined and communicated throughout the organization. Make sure that no obstruction, conflict or confusion may arise as a result of inadequate information given to corporate departments.

		Risk	Workpaper References
10.2.6 Coordination procedures for other relevant public authorities have been defined and documented.		**Medium**	
Security Concern/Implication If there are no coordination procedures and established mechanisms for institutions or organizations representing a legitimate public interest, the company under review may experience conflicts of interest during a crisis or disaster. Actions taken under the prerogative of public safety, health or other such interests may supersede corporate BCM activities and thus pose a significant risk to the corporate interest. **Recommendation** We recommend that, further to identifying all relevant organizations acting in the public interest, that clear and comprehensive coordination mechanisms be defined, documented and mutually agreed between the parties.	**Manual Review Procedure** Ascertain whether all persons, organizations, institutions etc. have been defined that might influence activities during a crisis or disaster. Verify if these organizations represent a legitimate public interest or not. Obtain and review all coordination procedures, approvals and memoranda and/or minutes of negotiations. **Automated Tool Reference**		

Detailed audit activities to be performed:

- Review the list of authorities, organizations etc. that have been categorised as "relevant" with regard to legitimate public interest. Depending on the geographic location of the audit, seek expert advice on the situation. It may be necessary to consider unusual scenarios, such as military intervention.
- Refer to section 2 of this audit programme to clarify if any third party intervention in the event of a disaster requires coordination or represents a significant risk. Assess procedures and actions that mitigate these possible risks.
- Collect and review in detail all documentation relating to coordination with authorities or parties exerting significant official influence on the BCM effort.
- If appropriate, highlight areas that may not be accessible to the auditor, such as confidential arrangements for executive protection or classified material. Ensure that responsibility for these audit items (which may be beyond the control of the entity audited) rests with the appropriate authorities and that this fact has been documented.

Overview Chapter 10 Audit Items

Number	Preliminary Rating Assigned by Auditor (from 1 to 5, 1 = absent or inadequate, 5 = comprehensive and adequate), and auditor's comment	Risk
10.1.1		High
10.1.2		High
10.2.1		High
10.2.4		High
10.2.5		High
10.2.2		Medium
10.2.3		Medium
10.2.6		Medium

11. Country-Specific Issues

11.1 Germany		Risk	Workpaper References
11.1.1 Business Control and Transparency Act (KonTraG) [Germany]		**Medium**	KonTraG, HGB § 317 (4), AktG, GmbHG
Security Concern/Implication If the business continuity concept as a whole does not address, or specifically exclude, the provisions of the BCTA, legal liability may be incurred concerning a) risk management and b) early warning systems. **Recommendation** It is strongly recommended that BCM as a business process be expanded to cover risk management and early warning requirements as mandated by the Act.	**Manual Review Procedure** Review the risks identified and possible activation scenarios contained in BCM / BCM documents. Assess whether an independent (extraneous) system exists for risk management. Assess links and cross-referencing between BCM / BCM and risk management procedures. Assess quantitative handling of risk and probabilities. Includes: Project risk management, investment planning, strategic decision analysis etc. **Automated Tool Reference**		

Detailed audit activities to be performed:

- Review existing documentation for risk management and risk analysis
- Assess integration with BCM / BCM and process interdependencies
- Compare current risk analysis procedures and mechanisms with BIA stage of BCM
- Examine risk mitigation elements in related areas - investment planning and budgeting, strategic decision analysis, project risk management

The KonTraG mandates that an "early warning system" be introduced to allow the management of the organization to identify any potential risks and events that might possibly endanger the existence of the entity (going concern assumption). The legal liabilities and remedies have been specificed in the AktG (Law on Public Limited Companies, specifically listed ones) and the GmbHG (Law on Private Limited Companies). Audit steps should cover both preventative and reactive components of such an early warning system.
For the purpose of auditing business continuity management, the reactive component is the part that should be reviewed. It is not necessary to review prevention, although deficiencies should be reported as findings outside the scope of the audit (there is a legal duty for the auditor to do so!).
Given that the legal ramifications of the BCTA / KonTraG are complex, it is highly advisable to use the services of a local auditor or attorney to ensure proper audit depth and completion.

			Risk	Workpaper References
	11.1.2 Banking Regulations (Bundesaufsichtsamt für das Kreditwesen) [Germany]		**Medium**	Bundesaufsichtsamt für das Kreditwesen: Richtlinien für das Betreiben von Handelsräumen
	Security Concern/Implication All banks or other financial institutions must conform with BA Kred regulations for trading and trading room recovery. **Recommendation**	**Manual Review Procedure** Obtain and review all BCM / BCM documentation for trading rooms and other trading locations. Check against BA Kred provisions and clearly distinguish German units from decentralized / cross border units. **Automated Tool Reference**		

Detailed audit activities to be performed:

- This audit item applies to banks or financial institutions regulated by the above-mentioned BA Kred procedures. It should be completed only for those clients who engage in trading activities

		Risk	Workpaper References
11.1.3 Other Regulations (Industry Sectors) [Germany]		**Medium**	Industry-specific regulations
Security Concern/Implication If industry-specific regulations are not reflected in the BCM / BCM, contradictory planning activities may occur. Misinterpretation and subsequent liability issues may arise. **Recommendation** Industries regulated by environmental or other regulations should incorporate the relevant parts in the BCM process.	**Manual Review Procedure** Review pertinent regulatory documents and assess the client's approach towards integrating them with BCM / BCM. **Automated Tool Reference**		

Detailed audit activities to be performed:

- Research should be directed towards identifying existing and impending legislation. While there may not be a direct regional or national document, European directives must be taken into account (delayed but mandatory introduction).
- Check environmental / health and safety regulatory context, including Berufsgenossenschafts-Richtlinien.
- Check (local!) provisions pertaining to disaster management and the role of public authorities - kommunale Katastrophenschutzverordnung etc. Contact local THW where appropriate. Contact local fire brigade where appropriate. Bear in mind that due to the federal structure of German lander, regulations may show some unexpected variation.
- If client is a) chemical, b) pharmaceutical, c) defense, d) power plant, examine in detail the relationship with local authorities, catastrophic disaster management plans, etc.

		Risk	Workpaper References
11.1.4 Fire Regulations / Building Safety [Germany]		**High**	GER: Feuerpoliz. Vorschriften, MBO, LBOen, BG-Richtlinien usw.
Security Concern/Implication If fire safety regulations are not complied with, events or critical situations might not be under control, and subsequent legal liability may arise. **Recommendation** It is strongly recommended that all fire safety regulations be complied with and updated at frequent intervals.	**Manual Review Procedure** Refer to appropriate manuals, legal documentation and other information sources for an independent review of factual regulations. Verify compliance with regulations; conduct interviews and walk-throughs. Check if premises have been inspected by fire brigade or other authorities. **Automated Tool Reference**		

Detailed audit activities to be performed:

- Obtain and review any inspection documents, including results and open issues. Valid inspections may have been performed by the fire brigade, the TUV, the VDA, the VdS (Verband der Schadenversicherer), internal auditors, or insurance auditors.
- Verify compliance with Musterbauordnung (MBO), Landesbauordnungen (LBO) etc., as documented.
- Conduct walk-throughs and verify if signage, escape room, fire fighting equipment etc. is in evidence. **Consult with experts** where appropriate.

	Risk	Workpaper References
11.1.5 Health and Safety at Work Regulations (Unfallverhütung, BG-Richtlinien) are being observed and integrated in the plans.	**High**	Berufsgenossensch. Richtlinien für die Unfallverhütung
Security Concern/Implication If BG Regulations are not observed, there is a significant risk of legal action following an incident, particularly so if persons have been injured or otherwise affected by the incident. **Recommendation** We strongly recommend that all BCM / BCM activities include BG regulations as an integrated part. **Manual Review Procedure** Review BG regulations (it is compulsory that these be displayed prominently on the premises) and check plans **Automated Tool Reference**		

Detailed audit activities to be performed:

- Check that BG regulations are displayed in a prominent location on the client's premises (this is a legal requirement).
- Verify that BG regulations are an explicit and integral part of all plans.
- Take into account that any accident or other impediment to a person or persons, as an independent event or in the course of a declared disaster, is subject to an external investigation by the BG. The plan should reflect this, as independent investigations may lead to serious legal consequences.

11.2 Australia, New Zealand		Risk	Workpaper References
11.2.1 ANZ 4360 Standard for Risk Management [Australia, New Zealand]		Medium	ANZ 4360
Security Concern/Implication While ANZ 4360 is not always binding, divergence from the standard can lead to misunderstandings, contradictory interpretation of risk management, and an insufficient degree of integration with BCM. **Recommendation** In client environments where ANZ 4360 is relevant, the standard should be incorporated into risk management and BCM.	**Manual Review Procedure** Address risk management procedures in direct discussions with management. Check control environment for ANZ 4360 compliance. **Automated Tool Reference**		

Detailed audit activities to be performed:

- Examine compliance requirements in accordance with the Standard, as follows:
- section 1: scope, application and definitions
- section 2: risk management requirements
- section 3: risk management overview (section 3.2: documentation must be sufficiently detailed to satisfy independent audit)
- section 4: risk management process
- section 5: documentation
- Assess integration of the risk management related activities (ANZ 4360) with business continuity Planning and management. Within the scope of this BCM / BCM audit, the focus is on risk management in the context of BCM. It is not an ANZ 4360 compliance audit.

11.3 Austria		Risk	Workpaper References
11.3.1 IKS regulations governing the internal control system [Austria]		**High**	
Security Concern/Implication If the internal control system does not address business continuity management, the operation of accounting systems and other vital functions may be impeded, thus leading to non-compliance with the law. **Recommendation** Business continuity management should be in line with the requirements relating to the internal control system (IKS).	**Manual Review Procedure** Review the relevant regulations. Check the BCM documentation for adequacy of internal controls. **Automated Tool Reference** N/A		

<table>
<tr><th colspan="2">11.4 Italy and Greece</th><th>Risk</th><th>Workpaper References</th></tr>
<tr><td colspan="2">11.4.1 D. Lgs. 626 / 94
[Italy and Greece]</td><td>High</td><td>D. Lgs 626 / 94
Dir. 89/391/EEC
Dir. 89/654/EEC
Dir. 89/656/EEC
Dir. 90/269/EEC
Dir. 90/270/EEC
Dir. 90/394/EEC
Dir. 90/679/EEC</td></tr>
<tr><td><u>Security Concern/Implication</u>
All provisions of this Italian law, as incorporating the abovementioned Directives of the European Union, have to be covered by a mandatory security plan (Piano di Sicurezza) in order to satisfy the legal requirements. The absence of such a plan poses the risk of non-compliance with Italian law.

<u>Recommendation</u>
It is strongly recommended that D. Lgs. 626/94 be incorporated as one central component of BCM inasmuch as its detailed provisions point towards safety and security.</td><td><u>Manual Review Procedure</u>
Obtain the Security Plan and perform a detailed analysis of its contents insofar as they relate to BCM and business continuity. Several areas of the security plan show some overlap or at least indicators as to the quality and completeness of the BCM.

<u>Automated Tool Reference</u></td><td></td><td></td></tr>
</table>

Detailed audit activities to be performed:

- check the risk impact analysis (business impact analysis) for each site. Focus on infrastructure, building safety, fire regulations and other risk-prone environmental factors which might become relevant in the event of a disaster.
- review the (mandatory) evacuation plan and its test results
- review the risk / reality gap analysis and the resulting action list
- peruse any logs and or minutes of unannounced checks in the past (by fire brigade or medical organizations)
- first point of contact is the assigned Security Officer on site.

		Risk	Workpaper References
11.4.2 Physical Security (Primary Facility) [Italy and Greece]		**Medium**	
Security Concern/Implication If the level of physical security around and within the primary site is not adequate, disruptions or disasters may occur which could otherwise have been prevented. **Recommendation** It is recommended that physical security be introduced as a significant preventive mechanism with a view to improving the effectiveness of the business continuity plan	**Manual Review Procedure** Review the premises and physical security concepts in place. Document by means of photography or other suitable actions. **Automated Tool Reference** digital or analog photography		

Detailed audit activities to be performed:

- check perimeter security (manned and unmanned), including single devices and integrated defences. Consult external experts where appropriate.
- check building outer shell and security concepts. Include passive and active infrared, cards and readers, mechanical security devices, etc.
- check internal security (buildings, zones, etc.)

		Risk	Workpaper References
11.5 United States and Canadian Standards on BCM and Risk Management			
11.5.1 US and Canada		**Medium**	
Security Concern/Implication as above, for clients being audited in one of the countries listed. Non-compliance with US or Canadian standards may have implications on other markets and subsidiaries of the client organization (cf. Y2K Readiness Disclosure Act). **Recommendation** Depending upon the specific provisions set out by the standard(s) applicable, compliance with these is strongly recommended.	**Manual Review Procedure** Research the background of standards and their international validity. Observations should include SEC / US GAAP related requirements whose direct implications are global. **Automated Tool Reference** N/A		

Detailed audit activities to be performed:

- Test client awareness with regard to these standards (interviews)
- Verify compliance (relevant sections only) by reviewing documentation
- Verify SEC / US GAAP compliance where appropriate
- Clearly delineate national vs. international consequences of standardization
- Ensure that BCM / BCM organization matches compliance requirements

Due to the fact that a comprehensive body of literature exists on both US and Canadian standards, the auditor is referred to these works for further reference. It may be advisable to distinguish between Federal, State and local regulations governing BCM.

Overview Chapter 11 Audit Items

Number	Preliminary Rating Assigned by Auditor (from 1 to 5, 1 = absent or inadequate, 5 = comprehensive and adequate), and auditor's comment	Risk
11.1.4		High
11.1.5		High
11.3.1		High
11.4.1		High
11.1.1		Medium
11.1.2		Medium
11.1.3		Medium
11.2.1		Medium
11.4.2		Medium
11.5.1		Medium

The auditor should bear in mind that these are suggested ratings. Chapter 11 contains regional and national specifics that may not always apply to the audit. In terms of assigning a risk factor, the above represents a preliminary rating. It has been inserted on the assumption that the English-speaking reader will be familiar with most US and Canadian regulations (hence the "medium," taking into account the auditor's expertise). However, this is not meant to imply that any of the local specifics listed in this chapter are comparatively less important or less essential to the functioning / compliance of a business continuity management process.

The auditor is strongly advised to adjust the preliminary risk ratings on the basis of which countries and regions should be covered by the audit plan. For reasons of space, not all pertinent regulations, laws and other directives have been included, and the above list does not claim to be exhaustive. It is recommended that the reader seek legal advice prior to acting upon the information provided, as has been indicated prominently and clearly in the accompanying explanations.

12. Software-Based Planning			
12.1 General Status		**Risk**	**Workpaper References**
12.1.1 Business Continuity Management Software is being used to plan and manage all activities.		**Low**	V. Roessing, Wiring (2000) BCM Tools Evaluation Paper.
Software is often used to streamline and improve the efficiency of business continuity management. The auditor should distinguish between dedicated BCM software, customised or self-developed solutions and other implementations of software-supported BCM.	**Manual Review Procedure** Review software as implemented. Review product descriptions. **Automated Tool Reference** Depending on product(s) used.		

Detailed audit activities to be performed:

- If software is being used to support BCM, assess the following:
- product name, description and references to manufacturer
- scope of software: full-scale tool, niche tool, secondary support software
- type of software: developed, bought customised, bought standard, Office customized

		Risk	Workpaper References
12.1.2 The BCM software package used is part of the corporate network and its access structure.		Medium	
Security Concern/Implication If BCM software does not form a part of the general corporate network structure, is may be isolated and therefore not easily accessible. As a result, its existence and use may be too restricted, and employees may experience difficulties with the software in the event of a crisis or disaster. **Recommendation** BCM software should be an integral part of the corporate network structure. Access to the software should follow normal access patterns in order to allow employees easy access and to encourage frequent use.	**Manual Review Procedure** **Automated Tool Reference**		

Detailed audit activities to be performed:

- Check if the software package has been installed alongside other user programs such as Microsoft Office. Verify that the server for BCM-related software and files is part of the standard network infrastructure.
- Manually test access from different locations throughout the enterprise being audited.

		Risk	Workpaper References
12.1.3 The BCM software used covers all countries and corporate entities that are listed in the BCM scope statement.		Low	
Security Concern/Implication If the software does not cover the agreed scope of the BCM process, countries and companies not under the umbrella of software-based BCM may not be able to follow certain instructions. Discrepancies with regard to planning, maintenance, and reporting may arise. In the event of a crisis or disaster, entities not covered by the software may not receive appropriate corporate support. **Recommendation** It is strongly recommended that all countries and/or companies be covered by the BCM software package used.	**Manual Review Procedure** Review the software installation and accreditation logs and document its deployment throughout the organization(s). Use in-built mapping functions where available. Obtain permission to conduct manual tests and review user / domain lists. **Automated Tool Reference** BCM software (in-built tools) Systems management software such as Microsoft SMS, Tivoli or CA Unicenter TNG		

Detailed audit activities to be performed:

- Interview IT personnel responsible for deploying and maintaining the BCM application. Obtain planning and roll-out documentation for the basic application software and subsequent customizations.
- Obtain and peruse software distribution logs (if automatic software distribution is in operation). Verify completion of roll-out.
- Check that all software deployment and update runs have been completed for all entities listed in the scope statement. Review user testing and approvals procedures where appropriate.

		Risk	Workpaper References
12.1.4 The software is licensed properly, and a service level agreement (SLA) exists. The software is updated on a regular basis.		High	
Security Concern/Implication If licensing is not comprehensive, infringements of copyright and license agreements may occur. If software maintenance and updates are not subject to appropriate service level agreements, reliability and availability of the BCM application may be compromised. **Recommendation** The BCM application should be properly licensed for all instances and entities listed in the scope statement. It is recommended that all licensing agreements include a basic license as well as appropriate service level agreements for maintenance and updates.	**Manual Review Procedure** Obtain and review licensing agreements for the BCM application. Check basic licensing, maintenance and update schedules. Verify that actions stipulated in the SLA have been carried out correctly. **Automated Tool Reference** N/A		Standard corporate licensing agreements Typical end-user licensing agreements SLA-related literature

Detailed audit activities to be performed:

- Map BCM application components, interfaces and required third party software (e.g. database engines such as Oracle).
- Obtain copies of licensing agreements for each of the mapped components. Refer to previous compliance audits as appropriate. Check the licensing structure (end-user, corporate, other).
- Check the agreements under review against standard licensing contracts for the country in which the audit takes place. Note any divergent or customer-specific modifications to the agreements.
- Compare the extent and coverage of documented licenses against the number of users, etc. in the field. Highlight any discrepancies in terms of over- or under-licensing and document the exact reasons.
- Review in detail the service levels agreed in contracts. Check against update and maintenance history. Focus on typical events representing active support requests and incidents leading to SLA invocation.
- Review the approvals and user acceptance documentation confirming that the SLA is being enforced and accepted by all parties. Obtain copies of recent update history and match against reality by taking appropriate samples.
- If necessary, interview randomly selected users of the BCM application to confirm that the package is functioning properly and in line with service expectations.

		Risk	Workpaper References
12.1.5 All business continuity personnel have been trained and certified for the software package used.		Medium	
Security Concern/Implication If BCM personnel have not been trained and certified for the software package used, there is a high error potential. As a result, planning may be ambiguous or deficient, and user error may be a predominant element invalidating plans. **Recommendation** It is recommended that all personnel using or administering the BCM software be trained appropriately. Training should be adequate to the tasks expected of the individual (administrator / user). We further recommend that certifications be included in the training plan in order to provide assurance for users and administrators.	**Manual Review Procedure** Review training plans and schedules for a period of at least one year prior to the audit. Ascertain the extent and perceived adequacy of the training. Refer to the software distributor or other external training firms offering relevant training courses to learn about "typical" training needed to properly understand the software package. **Automated Tool Reference** N/A		

Detailed audit activities to be performed:

- Request training plans and schedules for at least one year prior to the BCM audit. Obtain a list of training courses offered by the software distributor and third parties authorized by the distributor. Compare organizational training planning with the suggested training paths / certification paths for the product.
- Verify that within the entity being audited, personnel have been categorized according to their BCM function and the corresponding software use. Check that appropriate training schedules have been set up for technical users (e.g. administrators) as opposed to standard users.
- Take appropriate samples to ascertain that all users of the BCM software package are required and encouraged to keep abreast of developments within the product, such as major release changes or add-on features. Randomly sample training plans over several years to ascertain whether employees have had basic training and update training.

12.2 Technical Status		Risk	Workpaper References
12.2.1 All plans documented using the software package are updated in line with plan maintenance intervals.		High	
Security Concern/Implication If there is a time lag between scheduled BCM maintenance intervals and corresponding plan updates within the software, discrepancies in versions may occur. Individual department plans may not fit in with overall planning, resulting in incompatible plans or confusion about versions. **Recommendation** We recommend that all plans documented using the BCM software package should be updated and reviewed in line with general BCM maintenance intervals.	**Manual Review Procedure** Map the maintenance schedule against the last change date in the software. Highlight any delays or omissions. Use the inbuilt software audit trail where available. **Automated Tool Reference** BCM Software Package SQL query and report generators, such as Crystal Reports		

Detailed audit activities to be performed:

- Review the general BCM maintenance schedule for the organizational units audited. Note the target dates for changes or updates to become effective.
- Use the software audit trail (e.g., audit history database) to verify changes and updates, referring specifically to fields like "date last changed" or the running of update queries to a database. If the software supports customized reporting, create a report or filter that highlights discrepancies or delays.
- Investigate the reasons for delays, and document any technical and organizational difficulties that may have prevented compliance with predetermined maintenance intervals.

		Risk	Workpaper References
12.2.2 Plan printouts and other reports are readily available. All plan printout templates are documented.		**Medium**	
Security Concern/Implication If no plan printouts are available, accuracy and adequacy of plan templates may not have been tested. Undocumented printout and reporting templates may create confusion as to the facts being reported. Information reported by an unknown report template may be inaccurate or misleading. **Recommendation** We recommend that all planning templates be tested and verified by having at least one recent printout of actuals plans readily available. Non-standard printout (report) templates should be thoroughly documented and updated in line with maintenance or software update schedules.	**Manual Review Procedure** Collect and review plan printouts, matching the paper trail to the templates in use. Check documentation of templates, using a sampling approach. Perform in-depth checks where appropriate. **Automated Tool Reference** SQL query and report generators, e.g., Crystal Reports Database manipulation languages Standard high-level programming languages, e.g., Visual Basic		

Detailed audit activities to be performed:

- Collect a representative sample of plan printouts, taking into account criticality and BCM area covered. Ensure that at least one printout for each BCI area (e.g. emergency management group reports, BIA results overview, facilities management task list, general crisis command call tree, etc.) has been selected. Expand the sample if in doubt to maximize the level of assurance to be accomplished by this task.
- Map the printouts against the originating planning template. Distinguish between generated (e.g., MS Word), pre-formatted (e.g., MS Access internal reports), or externally generated (e.g. Crystal Reports engine) templates. Make sure that you understand the relationship between data source (i.e., the BCM database) and the queries that form the reporting template.

- If appropriate tools are available, select a random sample of templates and review the structure, particularly fields and SQL queries.
- As database analysis and the scrutiny of reporting templates may often prove difficult and technically complex, the integration of external experts (e.g., database administrators) is strongly recommended.
- Highlight discrepancies, questionable reports (which may yield inaccurate or misleading results) and technical errors or omissions. As the reports are the "end product" of a BCM tool, the actual output in terms of task lists, etc. is vital to the functioning of business continuity management.

			Risk	Workpaper References
	12.2.3 Software customizations are documented and known to business continuity personnel.		High	
	Security Concern/Implication If software customizations are not known to BC personnel, misunderstandings as to the functionality and setup of the software may result. This increases the risk of deficient planning. **Recommendation** All software customizations should be monitored and communicated in line with an adequate software development process.	**Manual Review Procedure** **Automated Tool Reference**		

		Risk	Workpaper References
12.2.4 Capacity management and continuous availability of the BC software have been ensured. Decentralized copies have been deployed and are replicated frequently.		Medium	
Security Concern/Implication If capacity and availability of the BC software are inadequately planned, this may create an additional risk of malfunctions in the software. As a result, planning data may be lost or inaccessible. **Recommendation** BC software should be subject to capacity management. It should be included in corporate high availability and IT disaster tolerance programs.	**Manual Review Procedure** **Automated Tool Reference**		

12.3 Software Functionality		Risk	Workpaper References
12.3.1 The software has an adequate set of functions to support overall business continuity planning and management.		**Medium**	
<u>Security Concern/Implication</u> If software functionality is restricted, there is a risk of incompatibility between software-based plans and other (paper) plans. **<u>Recommendation</u>** The software should contain all functions required for a complete BCM as defined by the organization.	**<u>Manual Review Procedure</u>** **<u>Automated Tool Reference</u>**		

		Risk	Workpaper References
12.3.2 Utilization of the software functionality in business continuity planning and management is efficient and comprehensive. The amount of manual workarounds is limited.		Medium	
Security Concern/Implication If manual workarounds exist to deal with certain weaknesses in the software, these increase the risk of software failures and mishandling of planning data. **Recommendation** Manual workarounds used to deal with perceived weaknesses in the software should be eliminated.	**Manual Review Procedure** **Automated Tool Reference**		

Overview Chapter 12 Audit Items

Number	Preliminary Rating Assigned by Auditor (from 1 to 5, 1 = absent or inadequate, 5 = comprehensive and adequate), and auditor's comment	Risk
12.1.4		High
12.2.1		High
12.2.3		High
12.1.2		Medium
12.1.5		Medium
12.2.2		Medium
12.2.4		Medium
12.3.1		Medium
12.3.2		Medium
12.1.1		Low
12.1.3		Low

APPENDIX A

SAMPLE AUDIT REPORT (FORMATTED)

Sample Audit Report (formatted)

Electronic Widgets Plc

Business Continuity Management Audit
as Part of the Due Diligence Audit for Electronic Widgets
Plc and German Widgets AG

February 2004

- Austria / Germany -

prepared by John Doe

1. Executive Summary

This audit report on business continuity management practices contains the findings, recommendations and analytical results for Austria and Germany. Audit activities were performed as part of an overall audit plans covering all subsidiaries of the Electronic Widgets Group of companies. This report is therefore part of a number of audit reports and opinions which should be viewed as a whole and within the scope of the overall audit objectives, as stated in the scope statements.

Goals and objectives for the audit of the Austrian and German subsidiaries are listed below. The findings and recommendations relating to these objectives were formulated on the basis of agreed audit procedures and international standards. The legislative and regulatory context have been included and indicated by references and remarks where appropriate.

- Compliance audit for all (Austria) or some (Germany) areas of the BCI / DRII model
- Due diligence audit with regard to a proposed merger between the Electronic Widgets Group and German Widgets AG, inasmuch as information about the latter is available

Findings

The *project management environment* for Austria and Germany is insufficiently documented and shows some methodological deficiencies. The business continuity management project preceding the transition to a full-scale management process does not focus on a clear scope and project plan. Several changes were made during the project lifecycle, some of which were occasioned by budget shifts and management intervention.

Risk evaluation and control activities have not been pursued at an adequate level of detail. The events and threats deemed relevant for the corporate entities in Austria and Germany reflect the results of several brainstorming meetings incorporating management and some specialist staff. Risks thus identified do not reflect the situation as expressed by published statistics and other independent sources. The risk evaluation process does not include the possibility of unforeseen events or major (catastrophic) events.

A formal *business impact analysis* has been conducted in certain parts of the Austrian and German organizations. However, the methodology and scope of the BIA do not cover core business processes in a comprehensive manner. Quantification of possible damages arising directly from adverse events has not been attempted. The qualitative assessment of direct and collateral damage is sketchy. Maximum tolerable downtimes and tolerable times in alternative operations have not been determined. As a prerequisite to the formulation of a BCM strategy, the level of BIA found at the time of audit cannot be regarded as sufficient or adequate.

Emergency response procedures and operations descriptions are well-defined in mandatory areas where Austrian and German law requires such plans. We discovered major weaknesses in those parts of emergency response planning that are not directly required by the law. Significant risks exist with regard to the ER and crisis management capability in both countries.

Austria and Germany have formulated a joint *business continuity strategy* on the basis of some business impact and emergency management considerations. The strategic focus appears to be on satisfying mandatory requirements at minimal cost, whilst consciously rejecting a comprehensive approach towards business continuity. The German subsidiary actively pursues an approach of diverse and limited projects for BC purposes, whose immediate objectives encompass legal changes and technical security. Core business processes are not covered by the BCM strategy.

Detailed business continuity planning for Austria presents a diversified picture. Well-documented plans exist for the more technical business processes, whereas core business has not been covered by any plans. The current functionality and plausibility of planning at a detailed level must be regarded as questionable.

Training and awareness activities at Electronic Widgets in Austria were conducted immediately after completing the initial cycle of BCM activities, i.e. on the basis of version 1 of the plans. No further training schedules have been set up, and the general awareness of staff (with the exception of IT personnel) appears to be unusually low.

Test planning and formal exercises of the BCM process have not been conducted. Some rudimentary testing took place in the context of Y2K crisis planning, but subsequent tests of updated plans were cancelled, due to budgetary pressure. Several parts of BCM have been inadvertently tested as a result of minor incidents in IT and facilities management. *Plan maintenance* has not been included as part of the BCM process. Some ad hoc maintenance activities do take place, but their occurrence is infrequent and unplanned. BCM cannot be considered up to date, and it is questionable whether Electronic Widgets staff might be able to enact the existing plans in the event of a disaster or crisis.

Public relations and crisis communications strategies have not been formulated in the Austrian environment, thus indicating a significant risk for Electronic Widgets. Senior management have not been included in activities aimed at developing a clear and coherent communications framework. While a public relations agency has been employed to deal with certain aspects of communicating risks and threats to Electronic Widgets, these efforts have not been integrated with BCM.

Coordination with public authorities in Austria is limited, which is partially due to the legal framework in which Electronic Widgets is operating. Some local arrangements have been put in place with the fire brigade and the local police. While this would not be acceptable in terms of BCM requirements, the regulatory situation does not allow further arrangements and activities on part of Electronic Widgets.

Recommendations

It is recommended that project initiation and management activities be centralized for Austria and Germany, as has been done with the strategy definition and documentation. This would reduce the risk of having inconsistent project management and control frameworks. Given the situation at the time of audit, it is further recommended that a standardized project management framework be applied to the current tasks and work packages as well as subsequent BCM development.

We recommend the introduction of a clear and comprehensive risk evaluation and mitigation methodology that should be based upon recognized standards. The current process of identifying and classifying risks by means of brainstorming meetings should be replaced with a more formal approach. This could be effected by including experts (for instance, the insurance department or external insurance representative), or by creating a risk-related management position within the organization. Given the nature of risks and threats in their local and regional economic environment, we emphasize that risk evaluation and control activities should be in line with these specific conditions, while adhering to any global standards in force within the Electronic Widgets group of companies.

We encourage both the Austrian and the German EW entities to continue their efforts with regard to assessing business impact in a formal manner. It is recommended that both entities build on the results achieved to date, and that they introduce a quantitative approach towards measuring and predicting possible damage that might arise out of critical events. Such steps should specifically refer to the internal controlling and management accounting framework, to identify and quantify any impact on the organization.

The emergency response and operations capability of EW in Austria should be extended and improved. We recommend that expert advice be sought to provide assistance in restructuring the crisis command center and its tasks. As the mandatory parts of emergency response have been planned at an acceptable level of detail, we suggest that this level of planning be seen as the minimum requirement for other parts of emergency response that have not yet been covered in the course of business continuity management.

With reference to the joint BCM strategy pursued by Austria and Germany, we strongly recommend that the "minimum cost" approach be reviewed and possibly replaced with a more proactive strategic posture. Given the risk / impact position that might apply to EW on a worst-case basis, it is evident that BCM requires strategic investments if it is to match both EW group standards and recognized BCM standards. We further recommend that the primary objective of formulating a BCM strategy should be the reduction of residual risk, and that the required investment be qualified in direct comparison with the possible damage arising from a crisis or disaster. The potential financial impact at the time of audit indicates a degree of exposure for EW Austria and Germany that cannot be classified as "residual", due to its magnitude.

It is recommended that the detailed planning area of BCM be extended to cover non-technical processes at a greater level of detail. We would encourage EW to bring all managerial processes to the level of planning found in most technical areas. It is further recommended that all plans be subject to a plausibility check. Such checks should be done in conjunction with a review of testing and maintenance (see below).

We recommend that training schedules and the general level of business continuity awareness be improved. This could be done by means of induction trainings, frequent awareness events and other forms of distributing BCM information. The current level of awareness among EW employees is significantly lower than average, which indicates that immediate attention should be directed towards informing employees about BC initiatives and planned steps.

Testing and maintenance of business continuity plans and management tasks should be improved and formalized on a short-term basis. We recommend that a comprehensive test schedule and corresponding maintenance concept be developed and implemented, to ensure the plausibility and workability of business continuity within EW.

It is recommended that public relations and crisis communications be integrated with the BCM framework. The current practice of having separate PR and BC plans cannot ensure that internal and external communications are planned and conducted in a controlled and targeted manner. We further recommend that specialist advice be sought where appropriate.

2. Audit Scope and Methodology

The review followed the framework set by the BCI / DRII Professional Practices, thus subdividing the relevant aspects of business continuity management into ten work areas:

- Project initiation and management
- Risk evaluation and control
- Business impact assessment
- Emergency response and operations
- Business continuity strategies
- Detailed business continuity planning
- Training and awareness
- Exercise and maintenance
- Public relations and communications
- Coordination with public authorities

While all ten areas were included in the case of Austria, the review was restricted to areas 1 to 3 for Germany. The restriction was imposed for reasons of time, and as a direct result of the self-assessment reports submitted by Austrian and German management at Electronic Widgets. Within the steering committee, it was decided that the level of BCM maturity reported by Germany did not require a full-scale audit.

A standardized set of questions was used as the basis for subsequent information gathering by means of personal interviews and questionnaires. Managerial and technical responses from EW personnel were tested by reviewing relevant documentation, by visual inspection (walkthroughs) and by means of direct physical testing of certain facts. The following table outlines the methods of information gathering for audit steps performed in Austria and Germany:

Area	Methods	Remarks
1 Project Initiation and Management	AUT: documentation review, interviews, cross-check of minutes and memoranda GER: documentation review, interviews, cross-check of minutes, memoranda and MS Project based files	project groups limited in size, personal interviews proved to be the most expedient means of gathering relevant information. Results of tests performed and copies of documents reference in work papers.
2 Risk Evaluation and Control	AUT: documentation review, interviews, review of published statistics and materials re. risks, interviews with leading insurance experts GER: documentation review,	external sources used for evaluation of plausibility of assumptions re. risk

Area	Methods	Remarks
	interviews, data gathering from published statistical sources	
3 Business Impact Assessment	AUT: process review, evaluation of computerized analysis tools, plausibility check GER: process review	no comprehensive BIA activity in Germany, audit steps therefore limited
4 Emergency Response and Operations	AUT: documentation review, critical assessment of past incidents and related records GER: assessment of past incidents and related records	both countries did not appear to have comprehensive ERO capabilities, although some mandatory parts exist *NB: GER included as a matter of convenience at no extra effort, since both firms pursued joint project on this*
5 BCM Strategy	AUT, GER: review of deductive process and final format used for formulating strategic options, viability, plausibility checks	*NB: both countries have joint strategy, GER therefore included as a matter of convenience at no extra effort*
6 Detailed BC Planning	AUT: documentation review, assessment and evaluation of MS Word based document management for BC plans	
7 Training and Awareness	AUT: process review	no training conducted, reviewed future plans only
8 Exercise and Maintenance	AUT: assessment of past incidents	no formal exercise and maintenance capability
9 PR / Comms	AUT: initial review	PR not included in concept and BC notion at the time of audit
10 Coordination with Public Authorities	AUT: initial review	Public authorities did not form part of the company's understanding of BC

As is often the case at Electronic Widgets, some areas of business continuity management are being addressed in a joint manner by Austria and Germany, both countries having German as their primary language. Some areas were therefore covered for Germany at no additional effort or expense.

3. Findings and Recommendations

3.1 Project Initiation and Management

The *project management environment* for Austria and Germany is insufficiently documented and shows some methodological deficiencies. The business continuity management project preceding the transition to a full-scale management process does not focus on a clear scope and project plan. Several changes were made during the project lifecycle, some of which were occasioned by budget shifts and management intervention.

3.1.1 Business Process Model

Some business processes in Austria and Germany are included in the BCM framework. However, some core business functions have not been addressed in the existing documentation. In the event of a crisis or disaster, delays and misunderstandings might occur with regard to recovery and restoration. This creates a significant business risk for EW, due to the possibility of prolonged disruptions and corresponding financial losses.

We recommend that the business continuity management concept be extended to all business functions and processes. The current gaps should be addressed by introducing a BCM model that mirrors the standardized EW process model. It is further recommended that the interdependencies between processes be identified and included in the BCM model.

3.1.2 Senior Management Sponsorship

Our interviews and further review of project documentation indicate a lack of senior management sponsorship. An individual manager or committee supervising the BCM project could not be identified. Project documentation shows that the BCM effort originated in planning for the "Y2K" project and was later continued under the heading of "Business Continuity Planning". No further support was given. The current project setup creates significant exposure for EW. While some investments have been authorized for perpetuating the "Y2K" type of continuity planning, the return on these investments is highly questionable. The gaps in the continuity concept (*see 3.1.1*) show that the lack of senior management sponsorship and the increased likelihood that core functions could not be recovered or restored in a critical situation.

We strongly recommend that all BCM activities be assigned to a person or group at officer level in order to ensure ongoing support and direction in the strategic sense. This could remove the current levels of uncertainty and the high degree of risk and exposure.

3.1.3 Project Management Lifecycle

The project management and control measures implemented at EW do not follow a defined pattern. Initiation, ongoing managerial control and reporting have not been formalized. Some ad hoc reporting requests and control meetings were held over a period of one year prior to the time of audit. However, the documents do not indicate that frequent and planned supervision of continuity management has taken place or is about to be introduced. The findings with regard to project management are reinforced by the lack of senior management sponsorship and the omission of several core processes.

The ongoing support and control of the business continuity project should be coordinated using a recognized project management methodology. We therefore recommend that planning, supervision and regular reporting be introduced to monitor the BCM project.

3.1.4 Business Continuity Coordinator

The position of business continuity coordinator is currently assigned on a part-time basis to a member of the facilities management unit who is reporting to the Head of facilities management. There is no direct reporting relationship to senior management. Responsibility for business continuity rests with the facilities management unit. As a result, there is inconsistency between the scope of BCM (covering the whole enterprise) and the span of control attributed to the responsible department (facilities management and technology). This inconsistency is evident in the planning bias towards technology processes in the current BCM concept.

We recommend that the position of business continuity coordinator be re-evaluated and given a direct reporting line to senior management. The current imbalance between technological BCM and the lack of planning for non-technology processes suggests that a member of the senior management group should supervise the strategic scope of business continuity and – where necessary – include additional personnel in the coordination effort.

3.1.5 BCM Budget

Activities related to business continuity management are not formalized by a separate and distinct budget. Funds for BCM are allocated on an ad hoc basis. Parts of technology planning have been included in the facilities management budget. For other business processes and departments, the department leaders have direct responsibility for authorizing BCM funds for steps and project phases suggested by the facilities management department. As a direct result of this split budget responsibility, there is no central control over the BCM budget. This represents a considerable risk to EW, as many departments have taken a reductionist approach towards business continuity.

The budget structure and split authorization principle should be reviewed. It is recommended that the final power of decision be assigned to a member of the senior management team to avoid conflicts of interest. We further recommend that the overall BCM budget be centralized and redefined to ensure adequate funding for the continuity process within EW Austria and Germany.

Item	Risk Weight	Rating (1 to 5)
1.1.1 Formal BCM	High (3)	3
1.1.2 Corporate Scope	High (3)	1
1.1.3 Introductory Info	Low (1)	3
1.1.4 Senior Mgmt Sponsorship	High (3)	4
1.1.5 Project life cycle	High (3)	1
1.2.1 Business continuity coordinator	High (3)	2
1.2.2 BCM resource assignment	High (3)	2
1.3.1 Separate BCM budget	Medium (2)	1
1.3.2 BCM separated from IT budget	High (3)	4
1.3.3 BIA alignment	High (3)	1
Average Weight, Total Rating	**High (2.7)**	**2.185**

3.2 Risk Evaluation and Control

Risk evaluation and control **activities have not been pursued at an adequate level**

of detail. The events and threats deemed relevant for the corporate entities in Austria and Germany reflect the results of several brainstorming meetings incorporating management and some specialist staff. Risks thus identified do not reflect the situation as expressed by published statistics and other independent sources. The risk evaluation process does not include the possibility of unforeseen events or major (catastrophic) events.

3.2.1 Potential Losses and Vulnerabilities, Risk Identification

While a general risk list was supplied for review, the process of risk identification cannot be regarded as methodical. Some brainstorming sessions were conducted in the past to create an initial catalog of risks and threats to EW Austria and EW Germany. The results, as documented in meeting minutes in separate documents for the Austrian and German entities, clearly show that the prioritization of perceived risks reflects the rank and seniority of the person suggesting the risk. In other words, the brainstorming meetings produced a list of fears discussed in a subjective manner. It is therefore difficult to assess the validity of the list of risks and threats. The documentation reviewed shows that only four brainstorming sessions were held, and no ongoing risk identification process is in evidence. Risks identified as a result of the ad hoc sessions early in the BCM project were not tested against published information regarding the probability of occurrence. The relative frequency of critical or catastrophic events in Austria and Germany do not form part of the risk identification and evaluation process. Business continuity management is strictly event-based, i.e. the planning does not envisage unforeseen events.

The situation found at the time of our review does not represent an adequate level of risk evaluation and control. The BCM concept must be questioned on the basis of this finding, as both the strategic and the tactical aspects of the BCM process rely upon the results of risk assessment and control.

We therefore recommend that the risk identification, evaluation and control mechanisms at EW be formalized as soon as possible. The suggested risk management framework should take into account any objective data as well as published information, in order to validate initial perceptions against known facts. The practice of attributing priorities to perceived threats in the order of corporate seniority should be changed immediately. While it is possible that some strategic risks are better known to senior management and therefore best expressed by these individuals, this should not lead to a situation where senior managers classify risks as valid or less important on the basis of subjective judgment, and without any corroborating evidence such as a BIA.

3.2.2 Prevention and Controls

Risk prevention and internal controls to ensure proper mitigation are in place for some business processes at EW Austria and Germany. The inherent bias towards technology processes (see above) has led to prevention steps that are almost exclusively aimed at reducing technical risk, whereas core business has not been addressed. The level of prevention is therefore at best uneven. Controls and mitigating steps are sketchy and do not follow a clear and methodical pattern. There is no integration with a general system of internal controls.

We strongly recommend that prevention be addressed in a more comprehensive manner. Mitigating controls and measures should be implemented methodically, covering all risks identified.

3.2.3 Ongoing Information Gathering

Our review covered the evaluation and organizational implementation of ongoing information gathering activities. A continuous approach towards obtaining updated information and data for risk analysis and control purposes has not been established. The assumptions governing the first version of risk and threat lists have not been verified in terms of changes in the natural and business environment.

It is therefore recommended that an ongoing process of information gathering be established to frequently update the business continuity knowledge base that has been built up within EW Austria and Germany.

3.2.4 Security Audit

Limited security audit activities have been performed at EW Austria and Germany. However, the documentation shows that these audits were designed to test specific security setups and products. Security-related risks as a distinct class of risks in general have not been identified.

We recommend the introduction of security audits in order to identify and evaluate security-related risks. These audits should include recommendations for improvements and mitigation. All risks should be integrated with the overall BCM framework.

Item	Risk Weight	Rating (1 to 5)
2.1.1 Potential losses, vulnerabilities	3	2
2.1.2 Prevention, controls	3	2
2.2.1 Analytical tools, methodology	3	2
2.2.2 Homogeneous risk analysis, regular intervals	3	1
2.2.3 Ongoing information gathering	3	1
2.3.1 Crisis and disaster scenarios	3	3
2.3.2 Risk classification, relevance	3	2
2.3.3 Potential impact on essential business factors	2	1
2.3.4 Security audit	2	2
Average Weight, Total Rating	**High (2.8)**	**1.8**

3.3 Business Impact Assessment

A formal *business impact analysis* has been conducted in certain parts of the Austrian and German organizations. However, the methodology and scope of the BIA do not cover core business processes in a comprehensive manner.

Quantification of possible damages arising directly from adverse events has not been attempted. The qualitative assessment of direct and collateral damage is sketchy. Maximum tolerable downtimes and tolerable times in alternative operations have not been determined. As a prerequisite to the formulation of a BCM strategy, the level of BIA found at the time of audit cannot be regarded as sufficient or adequate.

3.3.1 Business Process Priorities

While a business impact analysis has been conducted for parts of the EW organization, the review of related documents revealed that core business functions had not been analyzed. The impact assessment appears to be restricted to the processes that would be directly impacted by an event. Events causing a potential business impact are listed; the list reviewed did not cover unforeseen or indirect impacts on the EW organization. With reference to the findings in *section 3.2* of this report, this practice does not satisfy the criteria defined by recognized BCM standards: restricting impact assessment to some risks that have been identified without a methodology may lead to unforeseen impacts that have not been envisaged by management.

It is strongly recommended that the business processes and functions be prioritized in terms of criticality to the business of the EW group. The processes omitted in early impact analyses should be included as soon as possible. This is particularly important for core business processes that tend to be critical in terms of profit and loss.

3.3.2 Third Party Continuity

In the course of our audit, we could not confirm that all suppliers and vendors have BCM programs. A separate review of service level agreements and other contractual items revealed that EW in Austria and Germany has been granted a very favorable position with regard to products and services purchased from third parties. The legal framework in both countries supports these contractual arrangements. As a result, no further audit steps were taken.

3.3.3 Business Interruption Insurance

Neither EW Austria nor EW Germany have business interruption insurance policies. Given that no quantitative impact data could be obtained, it cannot be said with certainty whether the absence of insurance is adequate from a cost-benefit point of view.

We recommend that a detailed cost-benefit analysis be performed with regard to business interruption insurance. Insurance pricing should be taken into account and compared with the expected levels of damage as a function of time. This would enable EW to design a policy that covers either short-term or prolonged disruptions of core business.

3.3.4 Business Process Interdependencies

The business process model reviewed for Austria and Germany does not map interdependencies between processes. This represents a significant risk to EW, as cascading impacts or indirect disruptions of business processes cannot be predicted.

Process interdependencies should be identified and formalized using a standardized process model. It is important to identify the way in which processes are likely to be impacted following a crisis or disaster. The interdependency listing should therefore be designed in a way that gives a clear picture of possible cascading effects occurring as a result of process disruptions.

3.3.5 Maximum Tolerable Downtimes (MTD), Maximum Times in Alternative Operations (MTA)

MTD and MTA values based on the financial impact over time have not been identified. If these time values are not known, EW is unable to determine a strategy to recover and restore business processes in a controlled manner. Without MTD or MTA knowledge, the investment required to protect the business against financial losses cannot be determined.

We recommend that MTD and MTA values be calculated on the basis of potential impacts, business process criticality and process interdependencies.

Item	Risk Weight	Rating (1 to 5)
3.1 Comprehensive BIA	3	2
3.2 Prioritized business processes listed	2	2
3.3 Vendors and suppliers have BCM	2	2
3.4 Business interruption insurance	2	1
3.5 Business process interdependencies	3	1
3.6 Maximum tolerable downtimes	3	2
3.7 Max. times in alternative operations	3	1
Average Weight, Total Rating	**High (2.6)**	**1.556**

3.4 Emergency Response and Operations

Emergency response procedures **and operations descriptions are well-defined in mandatory areas where Austrian and German law requires such plans. We discovered major weaknesses in those parts of emergency response planning that are not directly required by the law. Significant risks exist with regard to the ER and crisis management capability in both countries.**

3.4.1 Emergency Operations Center

An emergency response and operations center (command center) has not been included in the BCM concept of EW Austria and Germany. Some organizational details have been defined with regard to the position of crisis coordinator and fire protection officer. There is no integrated organizational model that points to a distinct location exclusively used for crisis management or emergency operations. Our audit revealed that other criteria associated with the EOC have not been defined, although some sketchy descriptions exist about documentation of emergencies and collateral damage (injuries only). The fact that no central unit for coordination and emergency response exists increases the risk of confusion, misunderstandings and subsequent delays in recovering and restoring normal business. It cannot be ruled out that in a critical situation, parts of EW might be dysfunctional, without having the necessary support by a command center and informed

EOC personnel.

We recommend defining an EOC location with appropriate organizational and technical support. The selected location should take into account all requirements in terms of physical distance to the primary location, accessibility, infrastructure and other support issues.

3.4.2 Activation Procedures

Procedures for activating emergency response have been defined and formalized for those events that require a response strategy from a legal point of view (fire, hazmat incidents etc.). These event-oriented procedures are not linked by a unified activation scheme or chain of actions to be taken in the event of a crisis or disaster. As a result, there is a significant risk of uncoordinated actions and subsequent misunderstandings which might result in delays or deficiencies for the EOC.

We strongly recommend that activation, escalation procedures and the "taking charge" of the EOC be formalized, incorporating all isolated procedures and descriptions of emergency actions that exist within EW.

3.4.3 Containment Procedures

Detailed procedures for containing an incident or disaster have not been defined within the EW group. According to the teams and departments involved in managing specific events, this is due to the fact that the actual steps to be taken to contain an incident cannot be predicted with certainty. No further steps were therefore initiated. As a direct consequence, containment in a critical situation is likely to be based on improvisation and ad hoc measures decided by departmental managers. There is a risk of increased damage resulting from insufficient or delayed containment actions.

Damage assessment has not been formalized. There are no clear, documented methods governing the initial identification, subsequent evaluation and overall assessment of damage arising from a disastrous event or isolated incident. Hence, it might be difficult to ascertain material damage, should a sudden crisis or disaster occur. Subsequent substantiation of claims with insurers and third parties may be impeded by this fact.

We recommend that damage assessment, containment and related documentation be formalized at an adequate level of detail. Initial assessment of damage arising from an event should take into account the value of assets as well as intangible factors, such as danger to personnel or to the environment. The initial assessment of direct damage should trigger clear and concise containment actions, such as evacuating personnel or limiting a further escalation of the crisis. The following stages of damage assessment and estimation should be defined with a view to substantiating any claims against insurers or others.

3.4.4 Salvage and Restoration Procedures

There are no apparent links between the initial declaration of a critical situation or disaster, and the initiation of salvage and restoration. During our audit, we found that salvage activities were thought to be ad hoc and not related to priorities in terms of assets, although the safety of personnel had a clear priority in line with local regulations. Despite this fact, no clear guidelines and directives exist with regard to salvageable equipment or infrastructure. As part of our review, we examined the chain of events from the point of declaration to the hypothetical de-escalation of a disaster. It was found that no clear point in time had been defined at which restoration procedures are supposed to commence. This leads to an increased risk of delays in salvaging and restoring both equipment and business functionality.

It is recommended that, in addition to the mandatory components in conjunction with health and safety, the extent and plausibility of salvage activities be defined and documented. The current set of instructions clearly requires improvement, as there are no procedures for salvaging corporate assets. Furthermore, we recommend that the starting point for orderly restoration activities – following salvage, ongoing damage assessment and a preliminary damage report – be defined in the context of overall business continuity management. EW should bear in mind that an early start of coordinated restoration activities may significantly reduce collateral damage, and provide a favorable position in terms of subsequent insurance claims.

3.4.5 Expense Authorization

The span of control and authorization for expenses and other financial items is too narrow to adequately reflect the specific needs of a crisis or disaster situation. Expense authorization mechanisms have been designed in line with those under normal circumstances, i. e. in a very restrictive manner. While we refrain from comment as to this approach, we found that the comparatively limited decision making capability enforced by these limits is likely to impede emergency management, recovery and restoration activities. Furthermore, the number of managers authorized to sign off any expenditure is limited, thus creating an additional risk of delays and confusion in a crisis situation.

We recommend that the expense authorization mechanisms be improved for crisis and emergency management purposes. The overall limits and the persons authorized should be selected in the light of typical pressures accompanying a critical event.

Item	Risk Weight	Rating (1 to 5)
4.1.1 EOC / command center location	3	1
4.1.2 EOC and senior mgmt. needs	2	1
4.1.3 EOC procedures	3	1
4.1.4 Documentation, securing and logging evidence	3	2
4.1.5 Medical care, food etc. at EOC	1	1
4.2.1 Activate emergency response	2	3
4.2.2 Protection of personnel	3	5
4.2.3 Containment	2	1
4.2.4 Links to other BCM areas	3	2
4.2.5 Salvage and initiate restoration	3	1
4.2.6 Expense authorization	2	2
Average Weight, Total Rating	**High (2.45)**	**1.889**

3.5 Business Continuity Strategy

Austria and Germany have formulated a joint *business continuity strategy* on the basis of some business impact and emergency management considerations. The strategic focus appears to be on satisfying mandatory requirements at minimal cost, while consciously rejecting a comprehensive approach towards business continuity. The German subsidiary actively pursues an approach of diverse and limited projects for BC purposes, whose immediate objectives encompass legal changes and technical security. Core business processes are not covered by the BCM strategy.

3.5.1 Strategic Options

The predominant consideration in formulating a business continuity strategy for Austria and Germany has been cost minimization. As a result, a limited set of strategic options has been defined. We found that these options represent the bare minimum of continuity-related steps and concepts, without necessarily linking these to residual risk or potential damage. The reduction of BCM to a minimum cost exercise has led to a framework that does not address all necessary aspects of continuity. Existing strategic documentation is clearly restricted to isolated issues that are governed by laws and regulations. There is a probability that the current strategy, as selected from this set of options, may be dysfunctional.

It is strongly recommended that the strategic options be derived directly from the results of the business impact analysis. Any options identified and discussed should take into account both the maximum amount of potential damage to EW, and the residual risk after investing in prevention and recovery. We further recommend that the current strategy of minimizing any BCM investment be revisited with a view to targeted, increased spending in some areas where an unusually high risk of major damage to the companies exists and is clearly visible.

3.5.2 Technology and Non-Technology Issues

While the German subsidiary pursues an approach that covers some technical and some non-technical areas of the firm, the Austrian subsidiary has limited their BCM to cover technical items only. According to the information supplied at the time of audit, this is mainly due to the fact that the German Business Control and Transparency Act 1998 requires an early warning and risk control system spanning the whole business process model. Some weaknesses were discovered nevertheless: both in Austria and in Germany, the BCM is dominated by plans for technical environments such as facilities management and information technology. Several critical core business processes have not been included in the BCM strategy. In Austria, senior management have actively restricted the BCM project to cover technical processes only. We were informed that core business functions might be covered at a later stage, given the significant investment needed to extend the BCM concepts and planning effort. In our opinion, this constitutes a high-risk finding, as there is no apparent coverage of the critical functions in either subsidiary. It follows that – in the event of a disaster – non-technical processes may not be recovered in accordance with a set of plans. The success of core business recovery therefore hinges on improvisation and management awareness.

We strongly recommend that the core business processes in Austria and Germany be included in the BCM process and detailed planning effort. The necessary investment and resources should be assigned immediately in order to safeguard the critical functions that have not yet been covered by any plans and strategic thoughts.

3.5.3 Third Party Supplies

Third party products and services have not been rated in terms of criticality for the EW organization. As a result, there is no information as to whether the relevant suppliers have continuity plans in place. At the time of audit, we could not establish a definitive list of suppliers and service providers, as we were informed that some of the firms in the list provided to us no longer acted as business partners to EW. Products and services used by EW on a regular basis were thought to be covered by service level agreements containing legal remedies in the event of non-delivery. However, the case of supplier insolvency or inability to provide product or service had not been addressed by the purchasing department at EW. Further examination revealed that alternative suppliers had not been identified and contracted for several critical services. On the product side, some IT materials considered critical were found to be in stock at EW, but to a very limited extent.

It is recommended that any third party products or services deemed critical for the operation of EW's business functions be subject to continuity planning. This could be accomplished by integrating the relevant suppliers in EW's BCM framework. Alternative suppliers should be identified to avoid being dependent on a single company. We encourage the practice of holding in stock some critical products and spare parts. This should be extended to cover all areas of the firm, modeled on the basis of the current practice used in IT.

3.5.4 BIA Alignment

As a result of the weaknesses identified in the business impact assessment carried out in Austria and Germany, there is insufficient alignment between business impact and the strategic options. The viability of options, and their adequacy in terms of covering expected impacts and financial damage, has not been ensured. An alignment of strategic options with levels of business impact has been attempted in the shape of qualitative thresholds, by qualifying the options in terms of the required investment. However, without corresponding knowledge about the maximum damage caused by the impact of a crisis or disaster, the classification according to cost must be considered inadequate.

We recommend that the strategic options be subjected to a cost-benefit analysis, with specific reference to the results of a full-scale business impact assessment. The investment envisaged for each option should be compared with the maximum damage that might result from the residual risk tolerated when choosing this option. The level of residual risk presented by each option should be evaluated by management, prior to selecting one option that best fits the continuity strategy of the organization. It is further recommended that the business impact assessment steps be repeated in a more detailed manner where appropriate in order to avoid miscalculations.

3.5.5 Organizational Scope

The strategy does not cover all parts of the Austrian and German EW organizations. Some technical processes have been addressed in detail, whereas others have not been mentioned in the documentation. The core business units and processes have not been included in the definition of strategic options and continuity solutions. In terms of the joint strategy pursued by both countries, only the bare minimum of continuity-related strategic issues has been documented. The main body of documentation reviewed refers to legal requirements, particularly in Germany. Levels of responsibility throughout the organization have not been assigned in accordance with a comprehensive business continuity approach. There are some technical positions whose secondary function entails business continuity or "disaster recovery." However, there are no clear links to management that would indicate a senior managerial commitment and responsibility.

We recommend the introduction of a clear and unambiguous organizational model for BCM purposes. The continuity process should cover all organizational units and departments throughout Austria and Germany, including non-technical and administrative functions. We strongly recommend that particular emphasis be placed on the core business processes, such as sales and widget manufacturing, as these would suffer most from the impact of an unexpected critical event. Without including such processes in the continuity model, they may not be recovered and restored in an orderly manner.

Item	Risk Weight	Rating (1 to 5)
5.1.1 Strategic options	3	2
5.1.2 Technology and non-technology items	2	2
5.1.3 Third parties included in options	3	2
5.2.1 BIA alignment of options	2	1
5.2.2 Cost-benefit, viability, residual risk	2	1
5.3.1 Insourced / outsourced activity validation	n/a	n/a
5.3.2 Service level strategy for outsourced activities	n/a	n/a
5.4.1 Options cover whole organization	3	2
5.4.2 Options subdivide organizational units	2	3
5.4.3 Responsibilities at appropriate org. levels	2	2
Average Weight, Total Rating	**Med (2.375)**	**1.895**

3.6 Detailed Business Continuity Planning

Detailed business continuity planning **for Austria presents a diversified picture. Well-documented plans exist for the more technical business processes, whereas core business has not been covered by any plans. The current functionality and plausibility of planning at a detailed level must be regarded as questionable.**

Our review covered a representative sample of business continuity plans taken from randomly selected business processes in Austria and Germany. A list of plans was obtained to verify completeness of detailed planning in comparison with the overall scope of planning and implementing business continuity. We found that the majority of existing plans covered technical areas and processes, such as infrastructure, information technology and telecommunications. Other business functions, including the core processes that are vital to the organizations' value-creating activities, were either represented by sketchy and superficial plans, or not planned at all. Hence, the plausibility of detailed planning as a whole is highly questionable. While it cannot be denied that technical units will be able to apply their plans in a coherent and orderly manner, core business would not be recovered or restored, given the level of planning at the time of our review. The serious weaknesses in detailed planning for vital business functions is partially due to the fact that the overall scope of the business continuity project at EW did not specifically instruct process owners to embark upon planning activities. On reviewing relevant correspondence, it was found that the project as such had been presented to them as a technical exercise with little relevance to core business. It is nevertheless within the core process owners' responsibility to recognize the significance of BCM within the organization.

As part of the auditor's duty to report on practices deemed in conflict with existing laws, regulations or organizational duties, we state that the lack of planning caused by misleading correspondence might be construed as a lack of due care on part of senior managers and the board of directors. For EW Germany, there is no integration with the early warning system, and the provisions of the Business Control and Transparency Act 1998 may not have been satisfied by the current level of detailed planning.

It is therefore recommended that the issue be addressed immediately, by investigating the chain of events leading to this misunderstanding, and by taking appropriate corrective action. We further recommend that the organizational scope of BCM be expanded to cover all vital functions, in line with the duty of care to be exercised by departmental managers. Detailed planning should be initiated without any delay, including appropriate control and reporting mechanisms. The examples of technical plans reviewed should be used as an example for extending planning activities to other organizational units.

Item	Risk Weight	Rating (1 to 5)
6.1.1 Organization, teams, job descriptions	2	2
6.1.2 Planning aids, forms, information database	1	1
6.1.3 Comprehensive planning package distributed	1	1
6.2.1 Stages of escalation, trigger events	3	2
6.2.2 Escalation criteria defined, communicated	3	1
6.2.3 Recovery teams, roles, responsibilities	3	2
6.2.4 Clear distinction between recovery phases	2	3
6.3.1 Common planning structure	3	1
6.3.2 Comprehensive control and reporting system	2	1
6.3.3 Reference templates and samples for plans	2	4
6.4.1 Operations covered by detailed plans	3	1
6.4.2 Table of contents easy to use	1	4
6.4.3 Evacuation, fire, building safety	2	5

Item	Risk Weight	Rating (1 to 5)
6.4.4 Plan distribution list	1	1
6.4.5 Computing recovery requirements	1	5
6.4.6 Support equipment (technical)	2	5
6.4.7 Special forms and supplies (office)	2	4
6.4.8 Voice and data comms equipment	1	5
6.4.9 Backup tapes	1	5
6.4.10 List of personnel involved in recovery effort	2	1
6.4.11 Vendor / suppliers names and phone numbers	1	1
6.4.12 Team notification, call trees	2	1
6.4.13 Formal agreement recovery services / alternate site	1	3
6.4.14 Recovery site solution documented	1	3
6.4.15 Service levels for alternate site	1	4
6.4.16 Key contacts alternate site	1	4
6.4.17 Physical access controls alternate site	1	5
6.4.18 Logical access controls alternate site	1	5
6.4.19 Mail services alternate site	1	5
6.4.20 Salvage, cleanup, rebuilding of primary site	2	1
6.4.21 Manual fallback at 60 %	1	1
6.4.22 Alternate site appropriate	1	4
6.4.23 Offsite storage vital paper records	3	2
6.4.24 Third parties have adequate BCM (Detailed)	3	1
6.5.1 Plan completion monitored and validated	3	1
6.5.2 Planning glossary	2	1
6.5.3 All plans comply with planning guidelines	3	2
6.6.1 Distribution list satisfies “need to know”	3	3
6.6.2 Regular updates, version control	3	1
6.6.3 Third parties in distribution list	3	1
Average Weight, Total Rating	**Med (1.875)**	**2.227**

3.7 Training and Awareness

***Training and awareness* activities at Electronic Widgets in Austria were conducted immediately after completing the initial cycle of BCM activities, i.e. on the basis of version 1 of the plans. No further training schedules have been set up, and the general awareness of staff (with the exception of IT personnel) appears to be unusually low.**

At the time of our review, some three training courses had been conducted to present the first version of business continuity planning within EW. The ongoing nature of the continuity management process was not highlighted in training materials and presentation slides. While all departments were originally given the initial training, no further activities have taken place since. The level of awareness in non-technical departments is low, inasmuch as some employees remember the project, but others do not know about the requirements of business continuity within their departments. A random enquiry to employees revealed that over 50% of respondents confused BCM with the "Y2K" project and thought that no further activities were needed. Furthermore, some 15 % of new employees have joined EW since the introduction of version 1 of the business continuity concept. These employees are unaware of the fact that a BCM exists. A small number of respondents to the above-mentioned enquiry mentioned that the knowledge about the project had been lost, due to day-to-day pressure and other business priorities.

It is recommended that training and awareness events be conducted on a regular basis, and as part of employee induction procedures. The training materials should adequately address the fact that BCM is an ongoing process. Awareness materials should be distributed to employees to enable them to familiarize with the concepts and with detailed plans. It is further recommended that BCM-related activities be linked to corporate incentives to ensure active participation.

Item	Risk Weight	Rating (1 to 5)
7.1.1 External, industry, regulatory BCM support	1	3
7.2.1 Training BCM personnel	1	2
7.2.2 Other Staff trained	2	2
Average Weight, Total Rating	**Low 1.333)**	**2.25**

3.8 Exercise and Maintenance

***Test planning and formal exercises* of the BCM process have not been conducted. Some rudimentary testing took place in the context of Y2K crisis planning, but subsequent tests of updated plans were cancelled, due to budgetary pressure. Several parts of BCM have been inadvertently tested as a result of minor incidents in IT and facilities management. *Plan maintenance* has not been included as part of the BCM process. Some ad hoc maintenance activities do take place, but their occurrence is infrequent and unplanned. BCM cannot be considered up to date, and it is questionable whether Electronic Widgets staff might be able to enact the existing plans in the event of a disaster or crisis.**

There are no formal test plans and schedules for business continuity management. The Y2K test plans reviewed were found to be unsuitable for further use, as their contents and structure were too narrow. Upon detailed examination of correspondence and memoranda related to testing and further plan development, it was found that testing had been officially cancelled towards the end of 2000. Detailed plans, structural documentation and employee information, for instance in call trees, have not been updated or maintained since then. Random checks revealed that some employees who have left EW are still listed in continuity plans. We reviewed the notes and findings documented as a result of several minor incidents at EW over the past two years. None of the documents examined could be deemed appropriate in terms of the original plans and the minimum level of documentation described there. Isolated continuity plans have been updated on an infrequent basis following the aforesaid incidents, but such activities have been restricted to the IT department. Incidental memoranda and other corporate correspondence show that the main reason for discontinuing testing and maintenance were budgetary pressures and managerial instructions for cutting costs in several departments. It appears that BCM was one of the first items to be questioned and then removed from the departmental budget.

We recommend that EW introduce a detailed test and maintenance schedule based on regular exercises and defined maintenance cycles. BCM, as with any other product, requires constant updating and improvement to remain credible and workable within the organization. We further recommend that the budget for BCM-related testing and maintenance be assigned directly by senior management to relieve departments of undue pressure. BCM as a vital function that ensures corporate survival should be exempt from cost-cutting, as the removal of continuity funding immediately increases the risk of major damage in the event of a disaster. The money saved in the short term may well be offset by direct and collateral damage caused by dysfunctional plans and delayed recovery.

Item	Risk Weight	Rating (1 to 5)
8.1.1 Test results and deficiencies documented	3	1
8.1.2 Scheduled and unscheduled tests	2	1
8.1.3 Detailed test plans	2	1
8.1.4 Tests after major organizational changes	2	1
8.1.5 Regular and frequent (annual) testing	2	1
8.1.6 Tests address range of scenarios	2	1
8.2.1 BC plan version numbering	1	2
8.2.2 Documented procedures for updating plans	1	1
8.2.3 Responsibility for updating assigned	3	1
8.2.4 Plans stored onsite and offsite	2	1
8.2.5 Semi-annual updates	2	1
Average Weight, Total Rating	**Med (2)**	**1.045**

3.9 Public Relations and Crisis Communications

***Public relations and crisis communications* strategies have not been formulated in the Austrian environment, thus indicating a significant risk for Electronic Widgets. Senior management have not been included in activities aimed at developing a clear and coherent communications framework. While a public relations agency has been employed to deal with certain aspects of communicating risks and threats to Electronic Widgets, these efforts have not been integrated with BCM.**

We reviewed the general communications and PR strategy in force at EW Austria. It was found that there is no clear distinction between a normal state of operations and a crisis mode. According to information supplied by the marketing department, senior management did not deem it necessary to devise an alternative concept for crises or disaster, as none of these events had ever happened. Further interviews with senior managers revealed that they are confident about handling critical situations on an ad hoc basis, depending on the nature and extent of the event. The public relations agency stated that no specific instructions had been given to expand the general PR strategy to cover crisis communications. While the agency was aware of the needs and requirements of a crisis strategy, they had not been authorized to initiate such a project.

It is recommended that the current PR strategy be revised to include crisis communications. Senior management should determine the corporate position under adverse circumstances, and decide on a course of action to be accompanied by appropriate PR measures. It is suggested that the current relationship with the PR agency might be used to formulate a crisis communications strategy and implement detailed steps.

Item	Risk Weight	Rating (1 to 5)
9.1.1 Relevant PR components included in plan	3	2
9.1.2 External liaison points	3	2
9.1.3 Stakeholders and interest groups documented	3	3
9.2.1 Press releases and media statement	2	1
9.2.2 Other comms strategies	1	1
Average Weight, Total Rating	**High (2.6)**	**1.846**

3.10 Coordination with Public Authorities

***Coordination with public authorities* in Austria is limited, which is partially due to the legal framework in which Electronic Widgets is operating. Some local arrangements have been put in place with the fire brigade and the local police. While this would not be acceptable in terms of BCM requirements, the regulatory situation does not allow further arrangements and activities on part of Electronic Widgets.**

Our review focused on the existing relationships with public authorities. In contrast to the United States or the United Kingdom, the number of authorities or associations working in the public interest is limited. It is therefore difficult to establish liaison points with anyone other than the police and the fire brigade.

Without issuing a specific recommendation, we suggest that a more active participation in private sector associations and interest groups propagating BCM might be beneficial to the activities of EW. The findings in other sections of this report indicate that the free exchange of knowledge in these organizations may support the process of improving and maintaining BCM.

Item	Risk Weight	Rating (1 to 5)
10.1.1 Legal background thoroughly researched	3	4
10.1.2 Industry standards identified and considered	3	3
10.2.1 Statutory and official agencies	n/a	n/a
10.2.2 Liaison with DR / BC agencies	n/a	n/a
10.2.3 Local authorities	2	1
10.2.4 Police	3	3
10.2.5 Fire brigade	3	3
10.2.6 Other authorities	n/a	n/a
Average Weight, Total Rating	**High (2.8)**	**2.929**

4. Due Diligence BCM Opinion

Further to the audit activities at EW Austria and Germany, the review focused on comparing business continuity management as a whole with the situation of German Widgets AG. This evaluation was conducted as part of the overall due diligence process. The objective of comparing business continuity in both companies was to determine the levels of maturity respectively, and the required investment or possible benefits to the EW Group of companies.

As shown in the tabular representation below, the general level of BCM maturity at EW and GW is similar. The Austrian and German subsidiaries of GW are dominated by the local understanding of business continuity management, which tends to perceive BCM as a technical discipline restricted to IT, facilities management and infrastructure. EW, although not a continental European company, shows a similar tendency, as the level of BCM defined at group level has not been fully implemented for the local organizations.

4.1 Project Initiation and Management

The knowledge about project management capabilities at GW, particularly in the area of BCM, is limited. However, the current practice indicates that GW is following a standardized project management methodology which is likely to have been applied to the corporate BCM project. The levels of maturity for EW and GW are therefore similar. In the event of a merger, the EW investment required to standardize project management at GW is likely to be moderate. Conversely, the integration of GW's methodology within EW might be cumbersome in terms of time and effort.

Project management activities at EW and GW do not require a strict alignment when applied to business continuity management. In the event of a merger, a joint BCM project could be managed using different methodologies, provided that the scope and contents of the notion of BCM were defined in a comprehensive manner. Project management as a primarily administrative activity does not influence BCM as such, and it is expected that project administration could coexist, given the similar levels of BCM maturity at EW and GW.

4.2 Risk Evaluation and Control

GW have embarked upon a comprehensive project for identifying, evaluating and mitigating risk in a business sense. This is a direct result of the introduction of an early warning system to satisfy the criteria set down by the Business Control and Transparency Act 1998. As a German company, GW has been pursuing this project since 1999, and increased their investment in risk management in the course of Y2K preparations. The current level of risk management activities is therefore more advanced than the corresponding maturity level at EW.

In the event of a corporate merger, EW would benefit significantly from the mature risk management process at GW. Some assets, including the risk management tools developed by GW, might be integrated to enhance the speed and efficiency of EW's risk management.

4.3 Business Impact Assessment

Following the introduction of a risk management process, GW have elaborated their overall system of calculating potential damage to corporate entities and their supply chain. While these activities are not directly related to BCM, they nevertheless represent a solid method of risk controlling that could easily be extended to cover unexpected events and risks. The level of maturity is largely determined by the comparatively long experience that GW has accumulated in creating a BCTA-compatible risk management system. The business impact analysis method includes net present value calculations and decision support systems for investment and strategic decisions at group level.

A comparison with the business impact assessment findings at EW would indicate a substantial benefit in the event of a merger. EW is likely to increase efficiency and reduce the cost of conducting regular business impact assessments by using the GW processes that have been developed over several years. A moderate investment may be required to support the introduction of GW-style BIA activities throughout EW Austria and Germany. However, this investment is almost certainly lower than the cost of autonomous BIA improvement at EW (see recommendations above).

4.4 Emergency Response and Operations

The situation at EW and GW is identical, in that both groups have introduced emergency response and management concepts in line with legal and regulatory provisions. In the event of a merger, EW would be facing similar tasks internally and externally. The investment required to accomplish a satisfactory level of emergency response and operations is likely to be very similar for both corporations.

4.5 Business Continuity Strategy

The current BCM strategy at GW is not known by this name, and embedded in the corporate risk management process. It is therefore very extensive on the preventative side, although it does not contain clear links to the reactive parts of continuity management. Strategic options have been defined for certain investment decisions, thus providing a capability of calculating the relative value of various strategic alternatives.

Again, EW might benefit from acquiring the knowledge about strategic planning and evaluating strategic options in the event of a corporate merger. A moderate to low investment is likely to be required for adapting the methodology used at EW today to that found at GW.

4.6 Detailed Business Continuity Planning

Our information regarding detailed continuity plans at GW is limited, as these are considered confidential there. However, it appears that in a similar manner to EW, GW prepared a comprehensive set of plans for the Y2K transition. The quality of these plans is not known. Core business functions were covered to a limited extent, according to information supplied by GW.

It cannot be said with certainty whether EW would be able to derive any direct benefits from acquiring the planning method used at GW. However, the current state of planning at EW requires significant improvement and hence, substantial investments in BCM. It is therefore likely that, given the future development of BCM, a merger with GW would not lead to any significant increase in spending on detailed continuity planning.

4.7 Training and Awareness

EW and GW have pursued similar approaches with regard to BCM training and internal awareness campaigns. The Y2K transition period was used to create a coherent set of plans that were published and brought to the attention of all employees. Both corporations then reduced their efforts in terms of updating the BCM process and failed to maintain an acceptable level of awareness at staff level. In the event of a merger, it is likely that neither corporation would significantly benefit from the other as far as training and awareness are concerned. However, major restructuring and the immediate changes following a merger might be taken as an opportunity to reinforce BCM awareness and to initiate appropriate training activities.

4.8 Exercise and Maintenance

GW has introduced a strict and comprehensive test schedule for mandatory items such as fire drills. This is indicative of their preventative risk management approach. Testing does not appear to be frequent, and it does not include simulations or tabletop exercises for BCM. Plan maintenance is done for technical departments and processes, but not for core business.

EW and GW might benefit from a merger in that testing and plan updating would have to take place for the complete business continuity process. A significant investment is required for the general overhaul of plans following a merger or restructuring, regardless of the level of maturity of the two merging corporations. This investment is likely to be somewhat higher than the cost of autonomous improvement for testing and maintenance at EW.

4.9 Public Relations and Crisis Communications

GW has been involved in several incidents over the past years. Their public relations and communications strategy is not known in detail, but records and published materials indicate that GW has some experienced senior personnel whose PR capabilities are well above average. We could not establish whether an agency or other third party has been used to improve PR and communications for critical situations.

A corporate merger would enable EW to improve GW's concepts while integrating the skills and knowledge of the above-mentioned individuals. In conjunction with the agency contract, this would probably improve the general level of PR and communications efficiency. Cost savings can be expected but not quantified in this scenario.

4.10 Coordination with Public Authorities

Given that GW operates in a situation very similar to EW, the coordination issue can be regarded as secondary. There is no additional cost involved for either company on the assumption of an imminent merger. Public authorities in Austria and Germany do not encourage a participative approach (with the exception of the police and the fire brigade) for the type of BCM envisaged at EW. It is unlikely that coordinating BCM with the public interest would increase cost or BCM spending.

The following table outlines the strategic position of EW vis-à-vis a merger with GW. The maturity levels are shown in conjunction with the potential investment required by EW, or by both companies. Where the general maturity level is particularly low in an area of BCM, it has been assumed that EW would have to invest regardless of a corporate merger with GW. This is indicated by the word "both," meaning that the merged corporation would have to allocate funding to improvements in this area, without being able to benefit from existing arrangements or plans. The right-hand column indicates whether the acquisition of GW's BCM process is considered an asset (+ or ++), a neutral move (o) or a liability (- or --). It should be noted that these summary ratings reflect a somewhat subjective opinion, as is expressed by the estimates given for the maturity levels at GW. It cannot be stated with certainty whether these are accurate, given the limited access to GW data.

Area / Item	EW Mat.	GW Mat.	EW Invest	Overall
Project Initiation and Management	2.185	2 – 2.5	moderate	+
Risk Evaluation and Control	1.800	3	none	++
Business Impact Assessment	1.556	3	moderate	++
Emergency Response	1.889	1.5	none	o
Strategy	1.895	2.5	mod. / low	+
Detailed Planning	2.227	2 – 2.5	yes, both	o
Training and Awareness	2.250	1.5	yes, both	-
Exercise and Maintenance	1.045	1	moderate	-
PR and Communications	1.846	1	none	-
Coordination with Public Authorities	2.929	2.5	none	o

The table indicates that the general level of BCM is reasonably similar within both organizations. There are no areas that would require any disproportionately high investments by EW. From a BCM point of view, the merger may bring some benefits for EW, particularly in the area of risk management and risk evaluation.

APPENDIX B

SAMPLE WORK PAPERS (FORMATTED)

Sample Work Papers (formatted)

The following pages present a representative sample of work papers. These papers would be used by the auditor to make notes and substantiate the summarized findings in the report. While it is usually difficult to explain all details and relevant findings at length, due to restrictions imposed on report size and format, the work papers may be used as detailed, supporting evidence for disputed findings and recommendations. Where appropriate, footnotes may be inserted in the main report for cross-referencing purposes.

Depending on the overall agreement and project plan, some working papers may have to be presented to the auditee organization. It is therefore advisable to prepare them in line with the questions and audit program used for interviews, questionnaires and so forth. Work papers should be in a standardized format to allow easy filing and numbering.

Sample 1: Audit Item from Area 1 (Project Initiation and Management)

			Risk	Workpaper References
	1.3.2 BCM does not form a part of IT budget spending. Financial responsibility for IT-DRP and BCM does not reside solely with the IT manager.		High	
	<u>**Security Concern/Implication**</u> If BCM is defined as a part of IT spending, there is a significant risk of underfeeding. As a result, there is a high probability that BCM will be dysfunctional in the event of a disaster. <u>**Recommendation**</u> The budget for IT-DRP and BCM should not form part of the overall IT budget.	<u>**Manual Review Procedure**</u> Review IT budget. Check IT-DRP / BCM related items. <u>**Automated Tool Reference**</u> Financial accounting, MS Excel etc.		general financial budget IT budget BCM budget

Audit steps performed:

Reviewed copies of all budgets relating to IT and business continuity.

The general financial budget contains an IT item (not detailed). Further review of the IT budget showed that IT disaster recovery is included in the IT budget. No BCM items have been included. Cross-check with departmental budget revealed that "BCM" or "emergency management" items are part of department cost centers, not IT.

Documentation

- Sample copies of annual budget (stored in financial audit file and cross-referenced there)
- Copies of IT budget sheets, incl. MS Excel files
- Representative samples of departmental planning and budget sheets, sales, manufacturing, facilities management

References

A_AUSTRIA_20021231_IT Budget Excel
A_AUSTRIA_20021231_Sales Budget
A_AUSTRIA_20021231_Manufacturing Planning
A_AUSTRIA_20021231_FacMgmt

Sample 2: Audit Item from Area 2 (Risk Evaluation and Control)

		Risk	Workpaper References
2.2.2 Risk analyses are performed in a homogeneous and orderly manner, as well as at regular intervals.		High	
Security Concern/Implication If risk analyses are not performed frequently and in a uniform manner throughout the organization, unknown risks may exist. This in turn may lead to increased exposure and unknown vulnerabilities that may endanger the business. **Recommendation** The risk analysis process throughout the organization should be performed in a homogeneous, orderly manner. Regular intervals should be set for updating the results of the risk analysis and assessment.	**Manual Review Procedure** Review the documentation of previous risk analysis runs. Check that all organizational units are analysed on a regular basis. Compare results and highlight differences and discrepancies. **Automated Tool Reference** N/A		

Audit steps performed:

Interview with senior management

There is no full quantitative risk analysis. Interviewee felt that risks cannot be anticipated with a reasonable degree of certainty. No knowledge about recognized methodologies, referred to insurance dept. as the right place to ask.

Check on data gathering activities

No defined process for data gathering. No data analysis. Guesstimates are the rule rather than the exception. People rely on experience and history.

Verified documentation on past risk analyses

Sketchy documentation available for Y2K transition, but nothing more recent. Analysis then centered around the Y2K risk as such, which is a singular risk. No other risks were included.

Documentation

- Interview notes w/ senior managers
- Interview notes (manuscript) insurance department
- Copies of risk analysis documents for Y2K

References

no electronic files

Sample 3: Audit Item (complex) from Area 6 (Detailed Planning)

		Risk	Workpaper References
6.4.3 Emergency evacuation procedures, fire regulations, and other building safety procedures are included in the plan and tested.		**Medium**	High priority is assigned to the detailed shape and formal plan as mandated by standards and laws (see section 2)
Security Concern/Implication If there are no evacuation and safety procedures included in the plan, there is a potential for confusion during a disaster which could lead to personnel being injured or unable to escape. **Recommendation** We recommend including detailed emergency evacuation procedures, fire regulations, and other building procedures in the plan. These safety procedures are to be tested and updated as required.	**Manual Review Procedure** Query the appropriate party and/or parties to determine the necessary documentation to ensure safe evacuation and recovery for all personnel. Document the procedures in the plan if they are not present. If present, review for accuracy. **Automated Tool Reference** N/A		

Audit steps performed:

Obtained secondary information

General regulations for building safety, fire protection and emergency evacuation were obtained (in German) and translated. Double-checked with local fire brigade to avoid misunderstandings. Additional research by interview with local expert (consultant).

Walkthrough

Inspected all evacuation routes and other building safety measures with facilities manager. No weaknesses found except two fire extinguishers past maintenance date (were removed there and then to have it done the next day). Facility manager demonstrated sealed escape door after having informed the fire brigade, alert signal etc. worked fine.

Reviewed documentation

All documents listing fire drills and evacuation exercises were produced and reviewed. Fire drills done on a regular basis and signed off by senior management. No weaknesses in fire drills.

Documentation

- Regulations (in German), rough translation
- Work papers prepared by local consultant (in English)
- Copies of fire protection manual
- Copy of evacuation routes and building safety precautions
- "Brandschutzhandbuch" for corporate use
- "Organisationsanweisung Verhalten im Brandfall" procedures on what to do in case of fire
- "Gebäudesicherheit" security and safety manual
- "Richtlinien Berufsgenossenschaft: Unfallverhütung" mandatory health and safety manual that must be displayed prominently
- Consultant report (in German and English) on findings in conjunction with safety / fire and other mandatory items in Austria and Germany.

The local consultant used for reviewing the German manuals confirmed that this is standard practice and shows no deficiencies.

Sample 4: Audit Item (complex) from Area 7 (Training and Awareness)

<table>
<tr><th colspan="2"></th><th>Risk</th><th>Workpaper References</th></tr>
<tr><td colspan="2">7.1.1 BCM is supported by the industry, relevant associations and within the company. The legal, regulatory and cultural background reinforces BCM awareness.</td><td>Low</td><td></td></tr>
<tr><td><u>Security Concern/Implication</u>
If the economic and cultural environment of the company does not support BCM, there may be a discrepancy between internal and external perceptions. This may lead to a lack of credibility of the company as seen from the outside.

<u>Recommendation</u>
Internal training and BCM policies should be designed to adequately reflect the surrounding economic and cultural situation. Where a low priority environment for BCM is in evidence, the internal awareness program should reinforce the corporate vision of BCM</td><td><u>Manual Review Procedure</u>
Obtain an understanding of the cultural and economic background that may influence BCM in the country or region being audited. Compare corporate (internal) perception of BCM with the general understanding throughout the industry.

<u>Automated Tool Reference</u></td><td></td><td></td></tr>
</table>

Audit steps performed: (restricted, due to low priority of item and limited budget)

Legal and regulatory background

Reviewed legal statutes and some well-known cases to establish Austrian / German position regarding BCM. It appears that BCM as a concept is not well known and not usually mandated by law. However, technical and safety questions are addressed in a much more elaborate way than in the US or UK.

Industry background

The manufacturing industry is concerned with safety and security but not directly with BCM. Some indicators of BCM awareness are in evidence at factory level, where technical problems and difficulties are seen to cause direct financial damage. Managerial departments are less aware of BCM. The terminology tends to avoid BCM and call it "Sicherheit" (one word for security and safety).

BIBLIOGRAPHY

Includes works referenced in this book as well as further readings.

auf der Heide, E. Disaster Response: Principles of Preparation and Coordination. Republication of 1st ed (1989): Center of Excellence in Disaster Management and Humanitarian Assistance, 1999.

Australian National Audit Office. Business Continuity Management: Keeping the Wheels in Motion. Canberra: ANAO, 2000.

_____. Business Continuity Management Workbook: Guide to Effective Control. Canberra: ANAO, 2000.

Bank for International Settlements. Operational Risk. Consultative Documents. Basel: BIS, 1998 – 2001.

BSI. IT-Grundschutzhandbuch für den mittleren Schutzbedarf (IT Security Manual). Bundesamt für Sicherheit in der Informationstechnik, 2001.

CobIT. Control Objectives for IT and Related Technologies, v3.0. Rolling Meadows, IL: Information Systems Audit and Control Association (ISACA), 2001.

Craig, A. Business Continuity Management in Asia. Presentation given at the Survive US Group Meeting, Philadelphia 2000.

Delamarian, C., R. v. Rössing. A Connection Between Romanian Power Plants Using a Computerized Network. Mannheim, Timisoara: ISIM, 1997.

Emergency Management Australia. Non Stop Service: Continuity Management for Public Sector Agencies. Canberra: Commonwealth of Australia, 1997.

Garcia, H. F. Overcoming Typical Organizational Dysfunctions in Managing Crisis Communications. New York, NY: Clark & Weinstock, 2000.

Gesetz zur Kontrolle und Transparenz in Unternehmen (KonTraG) (Business Control and Transparency Act 1998). Bundesgesetzblatt, 1998.

Handelsgesetzbuch der Bundesrepublik Deutschland. Bundesgesetzblatt, 1998.

ISO / IEC 17799:2000. London: The British Standards Institute, 2000.

Kerzner, H. Project Management: **A Systems Approach to Planning, Scheduling, and Controlling**. John Wiley & Sons, Inc., 2000.

Levitt, A. Disaster Planning and Recovery: A Guide for Facility Professionals. New York, NY: Wiley, 1997.

PMI. A Guide to the Project Management Body of Knowledge (PMBOK). Newtown Square: Project Management Institute, 2000.

Propst, R., C. Dunkle, C. Beirne. Full Spectrum Incident Response Planning. Disaster Recovery Journal vol. 14, issue 1.

Rockover, G. Talking to the Media – When It's the Last Thing You Want To Do. Disaster Recovery Journal, vol. 14, issue 1.

Rothstein, P. Disaster Recovery Testing: Exercising Your Contingency Plan. Rothstein Associates Inc., 1998.

Rössing, R. v., M. Wiring. Tools for Business Continuity Management – Functional Framework and Criteria for a Software Based BCM Approach. The Business Continuity Institute, 2001.

Strohl Systems. Business Continuity Planning Guide. King of Prussia, PA: Strohl Systems, 1995.

ABOUT THE AUTHOR

Rolf von Roessing has extensive experience in business continuity management, information security and traditional security. Over the last years, he has worked with Ernst & Young in several European and global offices, including specialist assignments such as Y2K subject matter expert and active participation in several global core teams for business continuity. His current position includes BCM and security-related responsibilities, and he heads these service lines for Austria and several other countries.

Rolf is a board member of the Business Continuity Institute (BCI) and holds an MBCI certification. He is an active participant of the Institute's education committee, working towards integration of BCM best practices and tertiary education programs. These developments include the consolidation and publication of BCM knowledge, academic and research work.

In Austria, Rolf has contributed to several standardization and codification initiatives, notably the ISO 17799 introduction as a common standard throughout the country. He frequently supervises security-related certification examinations and has presented various lectures and training courses on business continuity management in a European context.

Rolf holds postgraduate degrees in Britain, France and Germany, as well as the CISA (Certified Information Systems Auditor) and CISSP (Certified Information Systems Security Professional) professional certifications.

"Auditing Business Continuity: Global Best Practices" is his first major book, following a solid background of academic publications and professional papers, including:

- "A Connection Between Romanian Power Plants Using a Computerized Network" (with C. Delamarian). Mannheim, Timisoara: ISIM (National Welding Institute), 1997
- 2001: "Tools for BCM - Functional Criteria and Framework for a Software Based BCM Approach" (with Michael Wiring). Published by the BCI.
- 2001: "BCM and Risk Management: Standardisation, Convergence and the Business Control and Transparency Act 1998."
- 2001: "Business Continuity Planning and its Framework of Reference: Laws, Standards and Integration With Insurance and Other Disciplines". Pune, India: Proceedings of the C D Deshmukh Joint Seminar on Information Systems Business Continuity Planning, National Insurance Academy of India.
- 2001: "High Availability: Just for eBusiness?" Presentation and Paper (in German). Leonberg (GER): Siemens AG
- 2002: "Security, eBanking, Digital Signatures: Potential Risks, Measures and an Integrated Audit Approach". Presentation and Paper (in German) for the Institute for International Research.

Rolf's educational background includes:

- Bachelor of Arts (Honours) European Business Administration, Middlesex University, London, upper second
- Diplom-Betriebswirt (FH), 1992 - ESB Reutlingen, Germany
- Master of Arts in International Business, DeMontfort University, Leicester UK, with distinction
- Diplôme d'Etudes Supérieures Spécialisées, Cadre Européen en Affaires Internationales, Université de Savoie, Chambéry, France. Mention bien (equivalent of Distinction)
- Fachzertifikat Internationales Management, 1993 - HTW Saarbruecken, Germany.

Rolf's professional background includes:

- 1996 – 2001 chapter board member, Project Management Institute, Frankfurt Germany
- 1993 – 1994 research on knowledge-based systems and neural networks in power plant engineering, wrote the commercial strategy for the European Union within consortium of 13 major utility companies from 9 countries
- 1994 – 1997 Head of IT Europe & Middle East Region, Securicor Security Services / Securicor Group
- 1996 – 1997 Romanian State Electricity Utility & Rom. government – planning and political proposal for nationwide power plant network for backup / remote control purposes, part of strategic initiative for a resilient power infrastructure in Romania
- 1997 – 2000 Ernst & Young Frankfurt, deputy head of European BCM services
- 1999 several large Y2K BCM projects
- 2000 – Ernst & Young Vienna, head of eSecurity Services, head of BCM for Austria, Croatia, Slovakia, Slovenia

OTHER BOOKS AND SOFTWARE AVAILABLE FROM

THE ROTHSTEIN CATALOG

ON DISASTER RECOVERY

www.rothstein.com

BUSINESS CONTINUITY: BEST PRACTICES

by Andrew Hiles

"**BUSINESS CONTINUITY: BEST PRACTICES** provides a practical implementation framework for the 10 core units of competence jointly established by the Disaster Recovery Institute international (DRII) and Business continuity Institute (BCI). It can be used as a step-by-step guide by those new to BC management or referred to by more seasoned professionals for ideas and updates on specific topics. The guide covers all units of competence decided upon by DRII and BCI, and adds further background based on the experience of the author. Exaples are provided throughout the book -- all having their roots in real cases. It is intended to be updated regularly."

- INFORMATION SECURITY MAGAZINE, August, 2000.

BCM FRAMEWORK™ CD: THE EASY-TO-USE, COMPREHENSIVE, CD-ROM-BASED BUSINESS CONTINUITY TOOL

by Andrew Hiles

BCM FRAMEWORK represents a new approach in business continuity management software tools. The problem with a lot of software is that it is expensive; it requires training in use of the software; it may require consultancy in its use; it does not necessarily match your culture, requirements, or documentation standards; and, it still requires a lot of work to complete the detail, without which any disaster recovery plan is worthless.

BCM FRAMEWORK is different. It is comprised of a number of easily tailored modules that are selected from our database of client work from a combined total of over one hundred years of consultancy experience - modules that are hand picked as the most relevant to your own situation, culture, organization, equipment platform and infrastructure. It contains Model Business Recovery Action Plans for key corporate functions with Organization Schematics and role descriptions, with some vital - and often forgotten - actions included.

ABOUT THE PUBLISHER

THE ROTHSTEIN CATALOG ON DISASTER RECOVERY

The Rothstein Catalog On Disaster Recovery has served as the industry's principal resource for 1,000+ books, software tools, videos and research reports, since 1989.

It is a division of **Rothstein Associates Inc.**, an international management consultancy focused on business continuity, crisis management, risk mitigation and disaster recovery, since 1985.

The Rothstein Catalog On Disaster Recovery is also the principal publisher in the fields of Business Continuity, Disaster Recovery, and Service Level Agreements.

A complimentary CD-ROM will be sent upon request to:
info@rothstein.com

THE ROTHSTEIN CATALOG ON DISASTER RECOVERY

www.rothstein.com
www.DisasterRecoveryBooks.com
www.ServiceLevelBooks.com
email info@rothstein.com